Julian John Révy

**Hydraulics of Great Rivers**

The Paraná, the Uruguay, and the La Plata Estuary

Julian John Révy

**Hydraulics of Great Rivers**
*The Paraná, the Uruguay, and the La Plata Estuary*

ISBN/EAN: 9783337232238

Printed in Europe, USA, Canada, Australia, Japan

Cover: Foto ©berggeist007 / pixelio.de

More available books at **www.hansebooks.com**

# HYDRAULICS OF GREAT RIVERS.

# E PARANÁ, THE URUGUAY,

## AND

# THE LA PLATA ESTUARY.

BY

## J. J. RÉVY,

MEMB. INS. C.E. VIENNA, ETC ETC ETC.

LONDON:

E. & F. N. SPON, 48, CHARING CROSS.

NEW YORK:

446, BROOME STREET.

1874.

# INTRODUCTION.

The largest rivers of the world **are on the Continent of South America.** Of these we surveyed the Paraná, **the Uruguay, and the La Plata; the former** produce the latter at their confluence. **The La Plata is, however, an estuary** of the South Atlantic, from **which the Great Rivers constantly displace the** sea-water.

At the time the Surveys **were undertaken we made special preparations to** ensure a degree of accuracy **generally not required for ordinary purposes, with** a view to collect a number **of reliable facts which might form a basis for** intellectual research on some **of the difficult and obscure questions of** Hydraulics.

The completion of the Surveys **had been inconveniently limited as to time;** and at their termination circumstances **intervened which postponed the con-** sideration of the Observations **for two years; it is only within** the last few months that we had an opportunity to **analyze** their result. **The** analysis disclosed the importance of the Observations; **and** it was not without hesitation that, after a full and elaborate inquiry, we determined to publish the Surveys. Their result is at variance with some of our accepted principles; and, as it is easier to undo the work of others than to replace it by a better structure, we hesitated to proceed. **To** avoid the difficulty, we have confined ourselves to the publication of the results and to their analysis, without particular reference to theories of great authors; and we offer none of our own.

It is a strange circumstance, that the Old World should have remained in comparative darkness on some hydraulic questions for the want of a truly great river, and that it should be through the instrumentality of a young rising nation that Europe receives enlightenment on delicate matters referring to the science of Hydraulics. We are indebted to the Argentine Confederation for the light which their giant river, THE PARANÁ, has so liberally spread over the movement of its waters; which little rivers shroud in obscurity, obliterating the traces which might lead to the discovery of the principles involved in their complicated movements. The Paraná will for the future occupy a prominent position not only among rivers, but also at our Academies and Technical Schools; and it cannot fail to bring the distant country before those who seek scientific education.

We might have greatly enlarged our work by the addition of numerous other Observations and their analysis at a variety of localities, such as: The Pavon, Ibicuy, Guazú, Obligado, &c. &c., Sections, and others on the La Plata; these, however, though they tend in the same direction, would not be instructive after the consideration and the analysis of the great Sections and Observations placed on record in the IIIrd, IVth, Vth, and VIth chapters. We withheld nothing which could have thrown light on the subject; we have not committed ourselves to a hypothesis or a theory; and we are content to place an historical statement before the members of our profession. The history of our Surveys may be useful to those who are accustomed to labour and to thought; and we venture to hope, that the appreciation of the result will do some good to all who will recognize the truth and the meaning of the following chapters.

J. J. RÉVY.

Hope Villa, Campbell Road, Croydon,
          December, 1873.

# CONTENTS.

## CHAPTER III.

# The La Plata.

## THE ESTUARY.

## CHAPTER IV.

## The Paraná de las Palmas.

### THE RIVER.

# CHAPTER V.

## The Paraná.

### THE RIVER.

## THE ROSARIO SECTION.

# CHAPTER VI.

## The Uruguay.

### FROM RIVER TO RIVER.

### THE RIVER.

## SALTO SECTION.

APPENDIX.

# APPENDIX.

# LIST OF PLATES.

## PLATE I.

**CHARTS OF THE GREAT RIVERS.** Chart No. 1, showing the La Plata Estuary called " Rio de la Plata "; the delta of the Paraná; the Guazú and Palmas branches; part of the Ibicuy and the Pavon channels; the locality of various sections; the main river Paraná as far as the town of the same name; part of the Uruguay lake. Chart No. 2. Detail survey of the delta of the Paraná, large scale; numerous branches of river; locality of sections; part of the La Plata and locality of current observations on the estuary.

## PLATE II.

**THE LA PLATA.** Diagrams of current and of tide observations; currents at surface; currents at a variety of depths below surface; bottom currents; curve of velocity of currents at surface and four feet below. Section of La Plata; velocity of movement of its waters at various depths at certain hours of the day; mean current observations. Inclination of surface from flood to ebb tide; minimum, mean, and maximum fall; tides during observation of currents; period of neap tides: range and variation of tidal wave.

## PLATE III.

**THE PALMAS AND THE LA PLATA.** Observations. Tidal rise and fall; spring tides on the La Plata; their effect on the Palmas; tidal wave, its imaginary and its true outline; relative position of the two systems of waves. The moon's meridian transits. Diagram of strength and direction of wind. Hourly surface current observations on the Palmas, day and night; curve illustrating increase and decrease of velocity of currents; dependence between the rise and fall of current curve and the rise and fall of river's surface.

## PLATE IV.

**THE PALMAS** (Paraná de las Palmas). Section of river; plan of survey; velocity curve of current across line of section; centres of gravity of section and of currents. Diagram of current observations; curve of surface current and corresponding rise and fall of river. Detail diagrams; velocity of currents, superficial and three feet below surface; mean, surface, and bottom currents.

## PLATE V.

**THE PARANÁ**; Rosario section. **THE URUGUAY**; Salto section. Outline of surface currents on line of sections; surface movement of rivers; centres of gravity of section and of currents; diagrams of current observations. Geometrical analysis of surface movement of rivers; velocity line of rivers; Paraná and Uruguay mean; inclination of rivers' surface. **THE MISSISSIPPI** at Vicksburg. **THE DANUBE** at Vienna. **THE THAMES** near London.

LIST OF PLATES.

## PLATE VI.

**THE PARANÁ NEAR ROSARIO.** Section of river, horizontal and **vertical scales 100 feet to** 1 inch. Outline of section and of surface currents; locality of observation **of depths and of currents.** **THE THAMES** at Thames Ditton, near London. Memoranda.

## PLATE VII.

**THE URUGUAY NEAR SALTO. Section** of river, horizontal and vertical scales 100 feet to 1 inch. Outline of section and of surface currents; locality of observation of depths and of currents. Memoranda.

## PLATE VIII.

**THE IMPROVED CURRENT METER,** and its application; elevation and section of meter, half full size; details; arrangements for the integration of currents from surface to bottom, and for surface current observations; disposition of boats on river for current observations of all kinds, and to any depth. Diagram of geometrical construction of value of the meter's records corresponding to the equation of meter.

# GREAT RIVERS.

## CHAPTER I.

THREE years ago we were requested to survey the La Plata, the Paraná, and the Uruguay in the Argentine Confederation of South America. The surveys were on Government account, as a preliminary step to large engineering works then in contemplation. Very little was known of these rivers at the time we undertook the survey, and the source of information deserving special attention was the British Admiralty.

We searched for other publications on the subject, and although we met with several good accounts on the resources of the Argentine Republic, they were comparatively of little value to the engineer, having been principally written to assist intending settlers, with a view to induce emigration to that country. The charts and books of the Admiralty were compiled from a variety of official sources, such as the French and the Spanish Admiralty, and also from the accounts of persons engaged in the exploration of the country. It was clear, however, that systematic observations on these rivers were never made. An officer of a gun-boat conducted a number of observations on tides and currents for a couple of weeks or months; five or ten years later another officer surveyed part of the coast, made soundings, and determined the outline of some banks; and so on, within the last thirty years a variety of information was collected by the Admiralty as the best known, and this was published for the use of navigators.

The Admiralty charts were our best guides. The La Plata appeared a large shallow estuary of the South Atlantic. The Paraná, a mighty river of great depth and width, surveyed from its mouth many hundred miles up river, showing undiminished size as far as surveyed; the upper portion of the river and the regions it traverses being unexplored, unknown. The Uruguay, also a great river, surveyed as far as the Great Falls near Salto, chiefly from old Spanish accounts, presented interesting features, which materially differ from the Paraná.

The survey of a river is usually a laborious work. With many of our small rivers it is often a tedious operation, and more so with large ones; traversing countries but little known and whole regions inhabited only by Indians. The requirements of a nautical and of an engineering survey also differ considerably. In the former, facts are collected referring to navigation, such as the outline and depth of channels, banks, currents, tides, winds, &c.; in the latter, we are concerned with the volume of water discharged in the varying conditions of the river, the nature of its bed, and the geological formation of the surrounding country; the currents are of greater importance to the engineer than the navigator; and accurate information on currents is perhaps the more important and the most difficult part of a river survey for engineering purposes.

From a careful consideration of the Admiralty charts we came to the conclusion that the survey of these rivers would almost entirely depend on new and systematic observations, and that the ordinary mode of proceeding would not be practicable. The La Plata is at its narrowest over 20 miles wide; the Paraná, over one mile; the Uruguay, over half a mile; and they are at many points much wider. The Paraná, discharging its volume into the La Plata by a number of branches, is surrounded by low-lying islands often submerged for a hundred miles, these forming the delta of the river. In such districts the use of theodolites, levels, chains, &c., is greatly restricted and often out of question, on account of the swampy nature of the ground covered with thick grass and jungle growing several yards high.

The mode of survey we adopted for these rivers was perhaps novel, and the plan we followed was probably not so much one of opinion or invention as one of necessity. We had to make the surveys ourselves without one trained assistant; and, with the instruments of our selection, we sailed alone to the field of operation. These circumstances originated with us as regards the sextant and the current-meter observations, since these instruments give observations independently of trained assistants. From the detailed description of the survey, which will form the subject of the next chapters, it will be seen how different it is to the ordinary mode of gauging rivers.

The survey of little rivers is plain enough, and, we may add, in many respects defective enough. Having selected a convenient locality, we stretch a chain or a rope from bank to bank at the intended section, and take soundings at measured distances, and we obtain the desired cross section of river. The current is usually measured by floats, carried about midway the river on the surface of the water, and the result of the observation is called superficial velocity of current. The fall is obtained by levelling a distance above and below the section, and these operations usually complete the gauging of the river and form the more important part of the survey. The data of the gauging afterwards pass through certain formulæ, the results of which guide the engineer in his views and opinions for his designs or reports.

Assuming the gauging adopted for little rivers to give satisfactory results, it is clear that every one of these operations will become more and more difficult as the river becomes larger and larger, until at certain dimensions they will be impracticable. We may stretch a rope across a river a hundred yards wide, but no one would attempt it with a river of a thousand yards width. A sounding rod may be convenient to a depth of ten feet, but no one would attempt sounding with a rod in a hundred feet of water; nor could a reliable sounding be obtained in the mode adopted for little rivers, by using the lead, the current would carry everything before it. As to current measurements in large rivers with the aid of floats, they become very difficult, and, we shall see, unreliable; and the fall of large rivers is usually so small, that—on account of surface undulation and local elevation arising from curvature of the bed or from the effect of wind—it may escape ordinary levelling operations. We may at much trouble level a long stretch of the river and get a considerable fall due to such a distance, but supposing the levelling to be most accurate with the best instruments and free from error, we only get the mean fall, which may or may not be the effective inclination at the section of the survey. This will hold equally with the fall so ascertained for little rivers, but not to the same degree, because with small rivers we need only level a short distance from the section and obtain a fall within the power and accuracy of the instrument, and it will be nearly the same at the section. Not so with large rivers; here the distance from the section must be great in order to get a perceptible fall, and the mean may differ considerably from that at the section. Indeed, with a river like the Paraná, levelling fails to disclose the fall, which comes within the margin of possible accuracy of our best modern levelling instruments.

To guard against failure in the gauging of the great rivers, we adopted many precautions which were considered unnecessary with ordinary rivers; and these precautions were fruitful in the discovery of new facts—remarkable for their beauty and simplicity, and which may open new fields for the thought of engineers. There are but few who take the trouble of measuring the currents of a river below the surface. It is so much more convenient to take the set formula of some distinguished author than to find the currents by observation; and those few who have taken some trouble in the matter, have rarely gone further than to observe a double float—two floats attached to each other—one at the surface and the other near the bottom, assuming that the movement of the double float will represent the mean current of the river, which may or may not be the case, there is nothing to show either way.

There are instances in which some observers displayed much ingenuity to measure mean currents by various contrivances, such as cylindrical rods reaching from the surface near to the bottom, and floating with the current. These attempts are interesting, although the results are of little value to the engineer. Our best authors made experiments with little artificial channels a couple of feet wide and deep, and so arranged certain formulæ that they closely agreed with the results of their trials; the formulæ were then extended to canals and rivers, and when a glaring discrepancy occurred between the adopted formula and an observation, a new "coefficient" was introduced to correct the erring formula.

Great mathematicians had considered the question of currents on abstract reasoning and argument, and making certain assumptions which formed the basis of their arguments, they determined the law of the velocity of current below the surface; and proceeding from different assumptions they arrived at very different conclusions. Thus, one author concludes that the current ought to decrease from surface to bottom as the ordinates of a parabola having a vertical axis and its apex at the bottom of the river. Another author concludes that the current ought to decrease from surface to bottom like the abscissæ of a parabola, with a horizontal axis, the vertex of the parabola being at the surface. These two conclusions differ as much from one another as if one authority advised us to keep our course true south; the other, that we should follow true east; they are 90 degrees apart. Others, again, derive the law of decrease to be elliptic, or one of the conic sections, and some thought it may be a straight line. These are obviously but as many speculations in the absence of positive information derived from observation.

The ingenious reasonings of the mathematicians were made the basis of experimental research by some distinguished engineers, who moulded the formula of the mathematician by the introduction of coefficients and constants until they closely agreed with the result of their experiments, and within the limit of their trials, or not far from it, these formulæ were satisfactory. Some acute engineers, however, found that the best formulæ were wide of the truth if applied to larger rivers, and that a change of coefficients was necessary to make calculation and observation somewhat agree; and so it arises that with different engineers we have different coefficients. The original formula may have an inherent defect in its construction when applied to the laws of nature; the defect may, within moderate limits, be neutralized by the artful application of coefficients and constants, but these never can eliminate the defect itself, which may be growing larger and larger as we depart from the little experimental channel and apply it to a river of moderate size only.

We have briefly alluded to the difficulties and uncertainties attending the survey of great rivers, and they are not only physical but also scientific, arising from imperfect knowledge and vague information on the subject. If the rules of hydraulic science in their application to European rivers become unreliable and uncertain, what may not be the result if the same rules be applied to rivers a hundred times the size of the largest in Europe? Physical difficulties may be overcome by perseverance and a little ingenuity; difficulties, however, arising from imperfect knowledge are of a different order. Under these circumstances we took the safe course, by assuming that we knew nothing about great rivers, and that we should have to obtain all information from observations upon the rivers themselves; and that we ought to be prepared to make them, assuming nothing and ignoring accepted rules. We also intended to make the observations with a degree of accuracy usually unnecessary for ordinary purposes, with a view to collect a number of reliable facts, which might form a basis for intellectual research in the discovery of the laws which govern the movement of water within confined channels.

We saw our way clearly to obtain without much trouble or difficulty, and without trained assistants, an exact section of the river, no matter what width or depth; and we settled the mode of survey in this respect in England. Not so in reference to current observations; these appeared by far the more difficult part of the problem. A plan had nevertheless to be adopted before we sailed, in order that the necessary instruments might be provided. We had a natural aversion to floats as a means to determine the currents of a river. It appeared to us a ready-rough way to observe currents. At first sight it seems to be a simple and natural thing to ascertain a current by immersing in it some solid matter which will float, and so trace the velocity and the direction of the current. Under favourable circumstances the velocity of the surface current may be observed by the movement of a float with considerable accuracy, and such results will be more reliable than those obtained from formulæ. Unless, however, the circumstances attending the observations be favourable in their integrity, float observations deserve no confidence. There must be a straight channel above and below the point of observation; the current must be steady and uniform, due to a long channel of uniform depth and width; there ought to be a perfect calm, or rather a gentle current of air of the same velocity and direction as the current of the river; and then, by observing the transit of the surface float between two sections, and by noting the time of passage, the velocity of current on the line of movement may be obtained with considerable accuracy. These conditions can, however, rarely be fulfilled, without which the results will be erroneous; and, in departing from the conditions, the most serious feature is, that we never know how near the truth or how far from it our observation may have been.

It is idle for some authors to maintain that the effect of wind on surface floats may be disregarded. If we place any float on the surface of a pond, and let it project ever so little above the water—and project it must, otherwise it could not float and could not be observed from a distance—the effect of a moderate breeze may be readily seen, carrying the float with it, in one minute a dozen or several dozen yards. What happens on the pond also happens on the river in a distorted form. The float will move the same distance in the direction of the breeze, and it will moreover be carried by the current another distance, and its movement will be compounded from the two. The superficial current of a river differs, however, with every foot of increased distance from shore; a float carried obliquely by the interference of wind will accordingly be carried by a variety of currents, and the float can give only some sort of a mean. The distance the float will travel in a given time will also be increased or decreased as the wind may be in favour or against the current.

These remarks on surface floats apply in an aggravated form to double floats to ascertain currents below the surface. We do not see how a float below the surface can, even approximately, give the current appertaining to a certain point of the section across a river; and if we speak of a current, it must be in reference to a definite point within the channel of the river, otherwise the expression signifies nothing. With double floats, two floats are attached to each other by a cord of a definite length; the one, being heavier than water, will sink below the surface; the other, being lighter than water, will float on the surface; and it may be made light enough to suspend the submerged float at a distance equal to the length of cord.

Let us first place this double float on a pond of greater depth than the length of cord. In calm weather the two floats will at once adjust themselves, the lower placing itself in a vertical line below the upper one, like a pendulum at rest. As soon as the upper float be moved in any direction, the equilibrium is disturbed, and the submerged float will have to follow in some fashion the movement of the upper one; and we know for certain that during the movement the lower one can never be in a vertical line below the upper float; that it must be some distance behind on the line of movement, and that the distance of the lower float from the surface must be less than the length of cord. The upper float may be moved fast enough to make the lower one rise to the surface; and this is one of the serious features attending the use of double floats—viz. the change of level by the lower one. How much the lower float will be behind the upper one, and how much it will approach the surface during its movement, will depend on its specific gravity and on the velocity of movement. The nearer its specific gravity will approach that of water, the more easily will it be displaced and held and remain in any position under water; and the smaller need be the velocity of the surface float to make the lower one rise to the surface.

All that has been said in reference to the movement of the double float on a pond, happens in a more complicated manner in the channel of a river. If the currents were everywhere the same, a double float would, in regard to the surrounding water, be in the same position as if placed on a pond, and the lower float would soon adjust itself in a vertical line below the upper one, and both would be moving with similar velocities down the river. Let us now consider the movements of the two floats down river, assuming for one moment different currents to impel the upper and the lower one. It does not matter which we assume to be the greater, the result would be the same. Having a different velocity of current, the floats would, if independent of each other, travel different distances in a given time. Suppose the upper float to be carried by a surface current of 300 feet per minute, and the lower by a current 200 feet per minute; then, if at one time the two floats happened to be in a vertical line, one minute later the upper float will have travelled a distance of 300 feet, the lower a distance of 200 feet. The two floats being, however, attached to each other by a cord of say 10 feet length, there will arise a dragging between the two of 100 feet per minute, the lower one retarding the movement of the upper, and the latter accelerating the movement of the lower; and their ultimate position will not only depend on the relative force of the two currents, but also on the sectional area of the floats and on their surface, on the specific gravity of the submerged float and the area and length of the connecting cord.

All the above quantities given, it would be a nice matter to determine the precise position of the two floats after a given time; one thing may, however, be taken for certain, the lower float could not at any time be in a vertical line below the upper, nor could the lower float move in a depth of water equal to the length of cord; and we apprehend, under all circumstances, considerable deviation from the vertical line and from the level on which the lower float was supposed to move.

To indicate the current within a depth equal to the length of cord, the resistance of the upper float must be nil, and the cord should possess no thickness, and the specific gravity of the lower float must be greater than that of water. These are the mathematical conditions to be fulfilled. It is obvious they can never entirely be satisfied, and we may consider how far a deviation from these conditions will affect the result. It appears in the first place that the conditions attached to each float are conflicting, and the more we try to fulfil those of the one, the more we violate those of the other; and that by compromising matters we must be content to obtain the mean current between the two floats, and so at least obtain something. If we were to make the lower float heavy, say double the specific gravity of water, the upper float would be considerably immersed, and, apart from its own weight, it would have to displace one-half the weight of the lower one. It would then be a nice matter for investigation how much the lower float with a given difference of current would remain behind the upper one, and from the inclination of the cord we might determine approximately the depth below the surface within which the lower float would be carried by the current. The form and size of the floats might be so adjusted, that with a difference of current of say 100 feet per minute, each would offer about the same resistance if moved with a velocity of 50 feet, the one on the surface, the other submerged; and then the upper one would be as much retarded as the lower accelerated; and by observing the velocity of the upper float we should obtain the velocity of the mean current between the two, and, as the superficial velocity might be observed by a single float, we might from the mean and the superficial current determine the current in which the lower one was carried ; and the depth within which it was moving having been ascertained, we should finally arrive at the velocity of current a certain depth below the surface.

If we attempt, however, so to arrange the floats, that the lower one should control the movement of the upper, so as to make the upper float indicate the movement and position of the lower one, failure is certain, and an indefinite number of complications arise. If the lower float is to control the movement and position of the upper one, the resistance of the latter must be insignificant, which necessitates a very small and light float with very little immersion. The lower float must be large, and offer much resistance, to alter as little as possible its position on account of some slight inevitable drag from the upper float and the connecting cord. So far everything might be arranged ; but the fatal part of the proposition arises with the specific gravity of the lower float, which must be nearly that of water ; that is to say, it must possess no weight when submerged, otherwise the little superficial float will be pulled down and disappear. It is easy to see what will happen with a large submerged float of no weight in water. It will float in any position and in any direction, horizontal and vertical ; it will be entirely at the mercy of the most gentle local current, which may carry it sideways, forward or upwards—the lower float may be anywhere within the range of the length of cord, and it will invariably be at a much higher level than that due to the length of cord ; accordingly, we do not even know to what depth the particular current we may have approximately measured will belong ; and observation ceases and speculation takes its place. As soon as we begin to compromise matters by making the lower float heavier, the upper one must get larger and more immersed, and then we can only obtain some sort of mean between the movements of the two floats.

We have hitherto considered the movement of the double float under favourable circumstances, such as a straight channel for a long distance, uniform width and depth, steady currents, calm weather, &c., &c., and we found their movements, even under the circumstances, anything but simple and certain. How the movement of the double float will be further affected by bends and irregularities of the channel, by currents having not only different strength but also different direction, by the effect of wind, &c., &c., we need not further inquire, because the problem would not only be too complicated for practical purposes, but it would be hopeless to extricate ourselves from the confusion arising from a number of contending and unknown forces operating on the double float. The difficulties, moreover, rapidly increase with an increased width and depth of river. Apart from the irregularity of the movement of double floats, the observation of the transit of a surface float a mile or so from the observer, becomes a difficult matter, the float appearing a mere speck on the surface, to be followed by telescopes and the second-hand of watches. Moreover, the lower float may be supposed to be 100 feet below the surface ; it may not, however, be as many inches below it. We may note the transit of the upper float through certain lines, and call the velocity obtained for it the velocity of the lower invisible float ; and say, that that velocity belongs to a current 100 feet below the surface, and carefully note the observations and the figures in the Survey Book. We believe this mode of proceeding to be on the principle of shutting the eyes before the difficulty, and of declining to see it.

We have already devoted more space to consider the movement of single and of double floats than they deserve at the hands of engineers, and we have done so because one of the largest of any river survey of modern times—the great Mississippi survey—had been conducted by single and double float observations upon a grand scale, as a means to determine the currents of the river. The engineers of that survey relied entirely on floats, and we consider it a misfortune to science and to practical engineering that so much ability, perseverance, and time should have been spent to obtain results which the unfortunate choice of floats has inconveniently marred and confused.

# CHAPTER II.

## ON THE MODE OF SURVEY ADOPTED FOR THE GREAT RIVERS.

---

### SELECTION OF LOCALITY.—SECTION OF RIVER.

WE think it will be to the purpose and to the point if we give a detailed description of the mode of survey adopted with one of the great sections, and take, as an example, the Paraná near Rosario; and also give an account of the selection of the locality for the survey of the river. A special steamer of small draught, with a numerous and good crew and several small boats, are indispensable for a prompt and satisfactory solution of the problem. The Government placed at our disposal the 'Aguila,' a small paddle-steamer of about four feet draught; three engineers who had just completed their academical education at Buenos Ayres; a captain and a pilot with a crew of about ten sailors; the steamer having in tow a large and several small boats, with a number of planks and poles, chains, ropes, &c., on board ship.

We sailed from Buenos Ayres on the 16th January, 1871, by the 'Aguila,' to survey the Paraná. We entered the Paraná de las Palmas, one of the main branches of the river; the other branch being the Paraná Guazú, about 40 miles to the north. It was our intention to gauge the Palmas and the Guazú, before they further divide into a number of smaller branches, and we had approximately determined the locality of the sections by a previous exploration of this part of the river.

The Palmas branch is from 400 to 500 yards wide, deep, with a strong current close to the banks, which on both sides of the river consist of low-lying islands of great length and width, submerged during floods. As a rule, the level of these islands is a couple of feet above ordinary water; they are covered with rank vegetation. We selected the locality of the Palmas section, and devoted several days and nights to the gauging of this branch, and a number of valuable observations had been obtained and placed on record in the survey books; these will be matter for another chapter on the "Paraná de las Palmas," and we merely refer to it at present as it happened on our way to another section, the details of which we are about to give. After several days' observation on the Palmas, we determined to abandon that section on account of the capricious interference of the tides of the La Plata with the currents of the Palmas, which forced us to seek for a suitable locality much higher up the river, to be out of reach and interference of tides.

The voyage up river continued at a slow pace, keeping a sharp look-out for a locality favourable for a section. Nothing suitable presented itself for a hundred miles up the main river, until the afternoon of the 22nd January we entered a straight reach presenting for

c

many miles a uniform appearance, bounded by a bluff on the right bank and by the usual low-lying island on the left. The course of the 'Aguila' was now changed, traversing the river at right angles to its current from one bank to the other about a mile apart, and navigable close to its margin; at the same time taking preliminary soundings at regular intervals of time, the speed of the steamer being reduced to about three miles an hour. These soundings, repeated on several lines some distance apart, disclosed a regular channel of uniform depth at similar distances from shore, varying from 5 to 71 feet of water. At the time we knew not how far this locality might have been from any town or village within miles, and we knew only from the captain that we were not far from Rosario. Having been already one week afloat with a small overcrowded steamer, we had to obtain a supply of provisions before we could engage upon further surveys. We proceeded accordingly to Rosario for the purpose, and in the evening made another excursion higher up the river to ascertain if within 10 or 12 miles an equally favourable locality could be found; but we met with nothing suitable, and returned to Rosario for the night.

Wherever the 'Aguila' was moored a gauge was immediately put down on the margin of the river, and hourly observations were made of the rise or fall of the level of the river. So at Rosario, although no section was here intended, the gauge was put down and observed. This precaution should never be omitted. The gauge did not vary a quarter of an inch in twenty-four hours, and it read the same in calm and a strong S.W. breeze. We were accordingly out of tidal reach, and a temporary breeze from that quarter did not affect the level of the Paraná; although later on, there was a heavy swell on the river, which on the 23rd made the little steamer roll inconveniently. In the afternoon of the 23rd we sailed from Rosario down the river to the locality which appeared favourable for observations and sections, about 12 miles below the town, and where preliminary soundings had been already taken the day before. In this reach of the river, the right bank is a vertical bluff about 70 feet high of rock formation; the left bank is a sandy beach, rising a couple of feet above the surface of the water, covered with thick and coarse vegetation and a few small trees.

At this time of the year the Paraná is usually rising, attaining high-water level during February, and maintaining that level with little variation during March, April, and May During our survey the river was in its ordinary low state, commencing to rise, gaining about one inch in twenty-four hours, the rise of the Paraná being three to four weeks later than usual. The gauge having been fixed at the margin of the river and the banks explored, it was found that the low-lying left bank was the more favourable for a long base-line, and accordingly a base of 3000 feet was measured by a steel tape of 300 feet in length; a peg being driven every length of tape, commencing from the line of section and running close to the margin of the river for the whole distance. At each end of the base a high pole with a large flag having been erected, another line at 90 degrees with the base at the intended section was determined by flags, the latter being the line of section across the river. This line at 90 degrees with the base could not be marked off with sufficient length on account of the swampy nature of the adjoining ground. A flag was accordingly moored upon a floating raft on the intended line of section about 800 feet from the base on the river, to mark the line with accuracy for observations of transit. A flag on the opposite shore, terminating the line of section, may also be put down, and may occasionally be of service. It is not essential that the line of section should be at 90 degrees with the base; but, apart from convenience, it is desirable that it should be at right angles, because the observation of the angle at the moment of transit becomes the most accurate and least liable to error, the angle then changing but little near the line of section; the change will be the more rapid the more the two lines deviate from a right.

Several prominent points on the right bank were then tied on the base-line by triangulation, observing the angle of the selected points with the base from both of its ends. These operations terminated the labours of the day on the 23rd, all being ready to determine the cross section of the river the next day, weather permitting. Accurate observations always require calm weather, and it is waste of time and of labour to attempt any on a windy day; and it is out of question to do anything in foggy or rainy weather. The distances are so great that the fixed points become uncertain and altogether invisible. A clear day with a gentle breeze is the best. With perfect calm it is often impossible to find the position of a flag if one or two miles distant, the pole itself not being visible without the waving red flag, which wants a little breeze to fan it and make it conspicuous. In a calm bright day observations on great rivers may be made with exquisite accuracy and ease if we are only fortunate enough to keep the flags in sight. With current observations perfect calm is preferable, and is the best; because, with the mode we adopted, transit observations are not necessary, and we may bide our time to find a drooping flag. Not so with observations for soundings; here the flags must be always clearly visible and followed, and the angle fixed at the moment of transit. A clear day with perfectly calm weather should be devoted to current observations. A clear day with a gentle breeze is the most suitable to determine the cross section of a river; other days are not fit for trustworthy observations; and from these remarks it follows, that the engineer should always try to make the most of favourable weather, which forms but a small percentage of the time which may be at his disposal.

The instrument of our choice for measuring angles was the sextant; and we found the little pocket sextant, giving angles to one or even half a minute, the most convenient instrument, and preferable to the larger marine sextants from six to eight inches radius. We have, with the pocket sextant under three inches diameter, measured thousands of angles during these great river surveys, all of which to the minute; and we have often in one minute's time measured and booked several angles. This is a truly invaluable instrument, and its use should be encouraged and extended in all engineering surveys. The theodolite was not required, and its use would have been greatly restricted and the exception. The larger kind of sextant, giving angles to ten seconds, was only used to check main triangulation points as obtained by the pocket sextant; and, occasionally, for latitude and longitude observation in conjunction with chronometers.

The captain and crew had orders to be ready early the 24th January, and steam was up by sunrise. The day turned out favourable, the strong breeze of the 23rd having ceased, and had given way to a gentle current of air. The lead was ready for sounding, having already ascertained by preliminary trials that the depth of water would not exceed 70 to 80 feet. We made it a rule not to take the depth of water by marks attached to the cord of the sounding-lead; we found, that the length so marked on the cord always changed, and in great depth the change amounted to several feet. When the lead touched the bottom a sailor took hold of the cord at the surface of the water and pulled it up; the cord was then laid by the assistance of others on the deck of the ship, and the length measured by a tape which was already stretched out for the purpose, the whole operation not taking more than one minute. From check observations we found the soundings accurate to the inch of depth. The captain having reported all ready, the steamer was allowed to drift below the line of section indicated by the flags already in position. On the bridge of the steamer, with sextant in hand, we directed the course of the vessel to cross the line of section near the intended point for a sounding to be taken. The steamer crossed the line of section at full speed, and proceeded, according to strength of current, from 150 to 200 yards higher up, when the signal was given to reverse the engines; and, the steamer was not only checked in its forward course, but was drifting astern with a speed equal to the current, which was readily ascertained by observing the water of the river and the steamer moving together without any relative velocity. At this moment the engines were stopped and the steamer was allowed to drift astern to cross the line of section. The sounding-lead was now lowered by a sailor close to the bottom, all drifting with the current together, and the soundings were taken as if on a pond; and before the line of section was crossed—say, within the length of the steamer—we ascertained the angle between the flag at the extreme end of the base and the two flags on the line of section; and, continuing to look through the sextant, the two flags on the line approached each other as we approached the section; and the moment they covered one another the flag at the far end of the base, already very near the other two in the sextant, was also made to coincide, so that the three flags covered each other, at which moment the signal was given to "sound," the lead being already within a few feet of the bottom. We then booked the angle observed by the sextant, and by the time we had done so the captain called out the depth in feet and inches, which having been noted at the observed angles, completed the operation. The engines were started again ahead, the line crossed a second time at any intended point, the engines reversed, the steamer drifted and crossed the line a second time, angle and sounding noted in the Survey Book, and so on, point after point was taken on the line of section with the corresponding depth of water, every observation being an independent one and entirely in the hands and under the control of one person.

An accurate and exact section of the greatest river may be obtained by the above method within a couple of hours. On the Great Paraná, the mightiest river yet surveyed, each observation for sounding appertaining to a definite point of the section, took, on the average of a number of observations from the left to the right bank over a width of river of more than 4800 feet, exactly eight minutes' time. It is, however, an essential condition that everything be perfectly arranged, otherwise it would be impracticable to get a single trust-worthy observation in a whole day. The observer must be "at home" with his sextant, he must be prepared, at a moment's notice and within a couple of seconds, to sight any two points within the range and power of his sextant; there is no time left for desultory attempts to "find" a flag and make it coincide with another, the time allowed for the observation of the angle is under ten seconds, and we found three seconds of time to be ample for the purpose. The angle recorded by the sextant must be read with the same promptitude as the time upon an ordinary watch, and always be read a second time as a check, after the first reading of the angle has been registered in the Survey Book; and the sounding obtained should be immediately noted to the angle before taking another observation. This is the easiest part of the operation, and the more difficult portion of the task is the directing of the course of the steamer and of the movements of the engines. A couple of hundred yards from shore the management of the engines is also easy, but as the distance increases, and with it the current, it is a difficult matter to estimate the direction of the steamer and the distance it had drifted below or proceeded above the section. After the completion of an observation proceeding to the next, the vessel in going ahead always crosses the line diagonally in order to get a certain distance away from the former point of observation; as soon as the line is crossed the course of the steamer is changed to nearly at right angles with the line. When the distance from shore becomes great, from 500 to 1000 yards, we are practically on a sea, and we have nothing to guide us for course or distance, not even time, because the strength of the current cannot be appreciated. Accordingly, the steamer may drift too far below the section, and in attempting to cross it diagonally we may get the next point of observation too far from the former or too near it—usually too far. Such a circumstance does not vitiate any observation, except that the soundings are on this account not uniformly distributed over the whole length of the section. The observer is, however, immediately aware of such circumstance by noting the angle of his instrument; it ought to decrease less and less as he departs more and more from the base, and at any irregularity in the decrease he should change the course of the steamer to get an observation at a greater or less angle, as the case may be.

The figures obtained for the soundings should also be watched. As long as they increase or decrease at a uniform rate, the course of the general proceeding need not be altered. Should, however, a sudden and unexpected change occur in the depth registered, the course should be altered, and soundings for intermediate angles between the two last ones obtained, to find the point of change in the bed of the river.

The crew of the steamer should be trained for the purpose, and a number of trial crossings should be made without noting observations in the Survey Book, the engineer merely training his crew. A good man in charge of the engines and another at the wheel are necessary. On the Paraná near Rosario, our crew was trained, having already taken a number of sections by a similar process. During observations no one was permitted to speak; and no order was given by word, all being done by signals, usually a sign from the observer's hand; and perfect silence reigned on board ship during observations, and all were at attention. Thus the great section of the Paraná was completed in two hours and sixteen minutes from the first to the last sounding.

The above description of the mode of proceeding adopted by ourselves for surveys to obtain the section of a great river, completes the observations in this respect, and the section may now be plotted, either by calculating each distance from the base by the observed angle, which would be equal to the length of base multiplied by the cotangent of the angle; or, by plotting the complementary angle—90 degrees minus the observed angle—at the far end of the base by means of a good protractor, giving angles to the minute. We usually adopt both systems, and check the one by the results of the other.

The number of soundings taken on a line of section may be multiplied to any extent by a repetition of the process described, and it will depend on the nature of the bottom of the river how far apart the soundings should be taken.  If there be a regular increase or decrease in several succeeding soundings, the bottom of the river may be taken of a uniform and even nature, and then the soundings may be a considerable distance apart, from one-tenth to one-twentieth the length of section; and a multiplication of intermediate points would be of no value, and could not sensibly affect the result.  If, however, the soundings are increasing and decreasing in succession, revealing an irregular bottom, they should be more numerous, in other words, nearer each other; especially if a great change in the level should be noted in two succeeding soundings, intermediate observations are necessary to determine the mode of change of bed.  In such cases the soundings may be double or treble the number, although we do not attach much importance to great numbers of soundings, and prefer half the number with double the accuracy of each observation.

## CURRENTS OF RIVER.

Current observations form by far the most difficult part of a river survey; it is a difficult matter with small rivers, and more so with large ones.  We determined the mode of current observation before we sailed from England to the River Plate, and provided ourselves with instruments for the purpose.  It is, however, easier to devise a plan than to carry it out successfully; and we had to make important modifications while engaged on the survey of the rivers before we could ensure satisfactory results.  We propose first to give a description of the plan we intended to follow, and of the instruments and apparatus with which we provided ourselves for the purpose; and subsequently a description of the modifications we adopted in consequence of experience gained.

In the preceding chapter we have given reasons at some length which determined us to abandon the idea of employing floats for current observations.  Currents ought to be measured and determined by special apparatus and machines.  A ship's log, for example, is such an apparatus, although of a crude conception.  A more refined arrangement, called current meter, was at the time constructed by leading mathematical instrument makers. We obtained one, and, with some modification, ordered two such meters for our observations. The design of these machines is ingenious.  A screw, similar in every respect to those employed for propelling vessels except size, having only four inches diameter, is made to revolve by the current, being in the foremost part of the apparatus and first met by the current.  On its axis there is a fine thread working in two worm-wheels of three inches diameter, each wheel containing a great number of teeth, and one of the wheels having one tooth more than the other.  The axis of the screw being held firm in the frame of the apparatus, these wheels are made to move the distance of one tooth for each revolution of the screw; and by the time one of these wheels has made a complete revolution corresponding to a couple of hundred of the screw, the second wheel moves in reference to the first the distance of one tooth, having one more than its companion wheel.  By the aid of the two worm-wheels, having divisions and indexes at their circumference, we are enabled promptly to read the number of revolutions the screw may have made, from which we may draw conclusions on the velocity of the current which made it revolve at a certain rate.

The principle on which this apparatus was made, is good; the mode however, by which we are to come to conclusions as to velocity of current, is defective. The apparatus was not designed for engineers. To make everything very easy, the manufacturers made the pitch of the screw equal to one foot. One wheel was divided into 252 divisions called feet, so that one complete revolution of this wheel was supposed to register a distance traversed by the screw of 252 feet; or, the apparatus being held in a fixed position, a similar distance traversed by the current. The second wheel was divided into 96 parts, and these divisions were supposed to represent furlongs and miles, so that by this instrument currents in the aggregate of 12 miles were supposed to be measured. All these attempts at making things very easy is a mistake, for they tend to make things very useless. The supposition that a screw of one-foot pitch will register a distance of one foot if moved under water, or if the water moved against the screw that distance, is in itself a mistake; nor could a screw of any pitch ensure, that one of its revolutions should be equivalent to a distance of one foot traversed by the current, if that screw is to measure different currents; because the value of one revolution of the screw as to distance will depend on the strength of the current; the percentage of slip would be much more with a gentle than with a strong current. Accordingly, the mode of "noting" the current as adapted for the instrument is erroneous in principle, and the result must be always wrong; and it will depend on the strength of the current whether the error is to be small or great, even if it had been correctly adjusted for a current of a certain velocity. To make the instrument tell the current is making too much of it; it cannot do more than register the number of revolutions of the screw, and that it may do with great accuracy.

There was no time to introduce material alterations, although some appeared desirable; and we had to content ourselves with minor modifications which could be accomplished within a few weeks. We ordered two new current meters on the same general construction, but only to register the number of revolutions of the screw, the diameter of which was increased to six inches to get a little more leverage to overcome the slight unavoidable friction of the mechanism. One of the wheels had 200 and the other 201 teeth, with a similar number of divisions on their circumference, by which arrangement we could measure 40,000 revolutions of the screw, leaving the determination of the value of each revolution as to distance with different currents to be ascertained by direct experiment. We recommended the utmost delicacy and accuracy in the execution of the meters, to be able correctly to measure gentle currents; strong ones are easily measured, even with coarse instruments. In practice we found these meters to give good results if carefully attended and cleaned, and that it would have been preferable so to arrange the pitch of the thread and the number of teeth in the wheels that 100 revolutions of the screw should correspond to a revolution of one wheel and 101 to the other, in order to make the divisions larger;—for, on account of the closeness of the divisions especially on the second wheel, and by the necessary play to ensure easy movement, it was sometimes doubtful which figure the index registered, and in such cases the observation was lost, because one division of the second wheel corresponded to 201 of the screw, making the result vague and for our purpose useless. So far we were provided with an instrument giving superficial currents with great accuracy, the value of each revolution having been first ascertained for currents of various velocities by direct experiment.

The next point was how to measure with this instrument currents in a considerable depth of water. The meter, attached to a rod, may be lowered several feet below the surface; but the pressure of the current, not many feet below surface, would make the observation doubtful on account of oscillations and vibrations, and the difficulty of managing the instrument. To meet this difficulty as best we could in the short time allowed by circumstances, wrought-iron tubing of about three inches diameter in lengths of about five feet was prepared, to be screwed together to any length required with a smooth joint outside. Our intention was to erect this tubing in the river on the line of section, and to hold it by moorings in a vertical position, the tube to be a guide in lowering or raising the meter any distance from surface to bottom. For this purpose a ring was attached between the meter and its directing vane, which having been slipped over the tube and attached to the meter and to the vane, kept the whole apparatus to the tube sliding up and down, and which was raised or lowered by cane rods, also screwed together for the purpose to any length required. To prevent the tube from sinking too much into the ground, a disk was attached a couple of feet from its end, resting on the bottom; and with a sufficient supply of tackle, ropes, chains, pulleys, &c., we were prepared for the expedition, and sailed from Southampton in October, 1870.

It was found in practice that the erection of the tube in a deep river with a strong current was a tedious and difficult operation. If provided with a sufficient number of barges and men and good tackle, the operation of mooring the tube might be safely accomplished in half a day, weather permitting. But these are precisely the things which, with an expedition on rivers but little explored, we can only command in a limited degree; there being neither men nor materials within reach to assist us. We erected the tube with considerable trouble in about 40 feet depth of water, the operation requiring nearly a day and all the hands at our command, including our own. In the evening after a hard and hot day's work, one of the mooring anchors slipped a little on account of the pressure of perhaps an increasing current, and in a moment the tube disappeared and was buried in the Paraná Guazú. This was a serious matter; we lost with the tube much of our tackle of ropes, chains, anchors, &c., and we immediately proceeded with all hands to pick it up. Before night set in, the tube was picked up and erected a second time, and having hauled the moorings tight we made safe that it should not fail again on that account.

The proceedings of this day impressed us strongly with the weak points of the mode we adopted for current observations below the surface of the water, and that it ought to be superseded by a better and more convenient plan. There was trouble enough to moor the tube in a depth of 40 feet, and it was easy to foresee what would be the difficulties in 80 or 120 feet depth, and a greatly increased current. The experience already gained seemed to settle the point, and we were satisfied that in the greater depths and the stronger currents of the Paraná failure would be almost certain; that we should have to abandon the tube for great depth, and if possible should altogether discontinue its use. These were highly inconvenient conclusions at the time. We were far away from help or assistance of any sort, in the midst of a wilderness of islands and swamps; not even a decent tree was within a hundred miles, and an expensive steamer with a full complement of crew and engineers waiting for orders and instructions. There was no help from without, it had to come from within. The harder the case and the more hopeless the position may appear, the sharper must be the device for a remedy, and usually the better will be the solution. Before sunrise next day a new plan was devised, and we could afford to look with indifference at the troublesome tube, which we left standing on the section with a large flag attached to it, to be the surprise of many a trading vessel navigating the Paraná, and a warning to engineers how not to do it.

We want to measure the current of the river by a meter at different depths from the surface of the water, say to 150 feet. It is therefore necessary to hold the instrument firmly at those distances from the surface, and if we can so hold it and set it in and out of gear at pleasure, the problem is solved. If no current existed, the meter might be simply lowered by a cord, having a weight attached to it in order that it may not be disturbed from its level by pulling a wire to put it in and out of gear. The current, however, carries meter and weight with it, and accordingly we want another cord in a horizontal direction to prevent the current from carrying the instrument away. The solution appears thus simple enough. A bar of iron, about two inches wide by ¼ inch thick and nine feet long, was taken and a hole drilled at each end. At the middle of the bar a short piece of round iron was attached to replace the wooden rod used for surface currents, and to this projection the meter was joined, allowing it to revolve as on a vertical spindle, the bar being held horizontally. To each end of the bar, cords were attached with marks one yard apart, similar to sounding lines. From a platform resting on two small boats moored on the line of section, the bar with meter attached was lowered by two sailors, one at each end; the men lowering the whole apparatus according to the marks on the cord, each to the same depth. Another boat was moored about a hundred yards higher up on the line of current, and from that boat a cord reached to the bar and was fastened to the hole in its end, which cord prevented the current from carrying the bar and meter with it, and compelled the whole apparatus to keep a certain distance, equal to the length of cord, from the boat moored ahead on the line of current. By raising and lowering the bar with the two cords attached at each end, the apparatus moves up and down in a vertical plane, and the cord operating in a horizontal sense from the boat moored a hundred yards ahead, keeps the apparatus a definite distance from that boat, and makes it move nearly in a vertical line. It is of no consequence whether the line be vertical or inclined, straight or curved; it is only essential that the apparatus should be held at the same level permanently during an observation; that it should not alter its position.

This simple arrangement was at once adopted and extemporized on board ship and tried, and it never once failed from the first trial to the last observation which we made on these great rivers; and it so happened that necessity and poverty of resources in a desert produced what abundance and wealth in the great Metropolis could not. We were no more limited by depth or by currents; the new apparatus worked with the same ease in 10 as in 100 feet of depth; in a current of one mile, or of five miles per hour; it was immaterial, we could practically go to any depth and observe any current.

But the new plan happened to ensure another invaluable result. We could by this mode of proceeding not only determine the current at a definite distance from the bottom with the same ease and accuracy as at the surface, we could also integrate all the currents from surface to bottom in a vertical plane, and so find the absolute mean of all the different

D

currents in a vertical line at the point of section under observation, and the result is mathematically correct as registered by the instrument, and independent of any argument or speculation; and we are only limited by the accuracy with which the meter performs its mechanical operation, which may at pleasure be increased to any degree. We have only to put the meter in gear, record the position of the indexes, and note the time of the commencement of the trial, viz. the instant it is immersed in the surface current; lower it at a uniform rate, raise it up again, and if we prefer, lower and raise it half-a-dozen times in succession, and note the time the instant it is raised out of the water, and again read the position of the indexes; by dividing the number of revolutions of the screw by the time observed, say by the number of minutes, if minute be the standard for comparison, and by converting the revolutions per minute into feet according to their value, we have the mean current in feet per minute with unerring certainty and remarkable accuracy.

We have often taken consecutive mean current observations, and found how with a slowly-increasing superficial current the mean current also increased with every succeeding observation by a fraction of a percentage, although one observation was taken say twice up and down, the other many times up and down from surface to bottom; each observation taking a very different time of minutes and seconds; yet, when worked out, the results indicated a corresponding gentle increase of mean current at the same locality and the same depth, with an increase of superficial current; the latter being independently observed during the time of mean current trials.

The apparatus as above described, in conjunction with the mode adopted for current observations from surface to bottom, we name "Current Integrator;" and in the Appendix will be found a detailed description of a meter resembling that which we used for the survey of these great rivers, with certain modifications and improvements which experience suggested; also a description of the mode of proceeding, and the arrangements to be adopted for current observations, which will generally ensure satisfactory results.

The engineer ought to be always perfectly acquainted with the power of his instruments, with their weak and strong points; otherwise an observation may be worse than none at all, because it may mislead. This will hold in a higher degree with instruments which during their operation are out of sight; and under these circumstances no observation is of much value, however carefully made, unless immediately checked by another similar trial. The result of the two observations should closely agree, and the engineer must judge how far they ought to coincide; if they agree within the margin of accuracy within which they may be made, the trial is good and should be immediately registered; if they disagree they should never be corrected, but registered as doubtful observations. These may be often of great value to detect adverse circumstances and forces not suspected at the time. The great and fundamental rule is, never to "improve" an observation, but to book it as it may have been obtained.

The danger attending the use of current meters is, that the instrument may be out of order, or its free and easy operation interfered with by accidental circumstances; a screw may be drawn a little too tight; sand or earthy matter clogging its wheels and increasing the friction, &c.—all of which tend to increase the slip of the screw and thereby make it to indicate a smaller current than the real one. The observer must be constantly on his guard, and watch the instrument with as much attention as the current which he is going to measure; he can in a moment satisfy himself whether his instrument is in perfect working condition, and this is to be done after each observation. By giving the screw a rotary motion by the hand, it ought to continue to run round like a fly-wheel, and a gentle breeze ought to maintain the revolutions of the screw any length of time. Should it decline to do so, the instrument must be at once attended to, usually cleaned and oiled.

For surface current observations, a boat should be anchored fore and aft in the current to prevent it swinging about; the position of the indexes of the instrument is then read and booked, and a watch with independent seconds held by an assistant before the observer. For such trials the meter is usually put into gear before its immersion into the current. The meter should be immersed near the bow, a couple of feet from the boat, in order to make the full current operate upon it. We proceeded usually as follows:—The meter being held close to the surface of the water as the second-hand of the watch was approaching the full minute, we have, at the instant the full minute was reached, dipped the instrument about six inches below the surface, and held it in that position one minute; the instant the minute was completed the meter was raised out of the water and an assistant was in readiness to check the screw as it appeared above the surface, which would have continued to revolve several times more if not checked. The position of the indexes being read and registered, and the exact time corresponding for the observation being noted, the minute observation was completed. Immediately followed a similar trial, the only difference being that the meter was exposed to the current five minutes instead of one, as in the preceding trial, so that the same current was invariably measured by at least two trials, one as a check upon the other. Each observation was in every respect independent of the other, and this proceeding was followed whenever practicable. The five-minute observation was the one by which we determined the current in subsequent investigations, and the minute observation was used as a check upon the accuracy of the observer and that of the instrument. The results usually agree within one per cent.

Where great accuracy was desirable, and in all observations of currents a considerable distance below the surface, the proceeding was somewhat different. With these the meter was thrown out of gear and submerged and placed in position on the desired level, be it one or a hundred feet below surface; and by means of a wire reaching from the meter to the observer, it was at a given moment thrown into, and at the completion of the trial thrown out of gear; all of which was done by pulling a wire tight at one time and letting it go at another. Here, again, always two observations were made on the same level, the meter being brought after each to the surface and the position of the indexes read. It will be noted, that during these observations the screw continued revolving, but being out of gear it did not turn the wheels, and consequently did not register revolutions until the wire was pulled, and it registered only as long as the wire was held tight by the observer. Usually a minute observation was taken first, to be followed by a five-minute observation, and the results compared before another was made. It is hardly necessary to say, that the observations are often not exactly of one or of

D 2

five minutes' duration, but they may be several seconds more or less, all of which is of no consequence as long as the exact time be noted in minutes and seconds; the one may be 59 seconds and the other five minutes three seconds, or any other figure; it is only for convenience that we attempt full minutes.

For mean current observation, namely, to integrate all the various currents from surface to bottom, the meter is placed into gear before the commencement of the trial, and no wire is used. All being ready, the time of immersion is noted, and the meter uniformly lowered by paying out equal lengths of cord at each end of the horizontal bar used for current integration, until the bottom be reached within an inch or two of the screw blade. To ensure that the meter be not lowered too far, and yet very closely to the bottom, a disk was attached to each end of the bar, the surface of these weights being a couple of inches in advance of the meter. As soon as these disks touch the bottom it is felt by those who lower the apparatus, and they at once begin to raise it again to the surface. In these trials the time is always several minutes, and no attempt should be made to make it full minutes, because the movement should be uniform throughout; and, whenever the surface be reached, the time should be noted in minutes and seconds. It will be always preferable to lower and raise the meter several times in succession to equalize any irregularity in the speed of movement. The time should only be noted at the commencement and the termination of the trial. The cords, at each end of the bar, should be marked red and white in succession every three feet, and the sailors engaged in lowering have to call out the colour of the mark reached, both of which must be immersed simultaneously; this will ensure the bar being lowered or raised horizontally, otherwise it would soon be inclined considerably, and with it the axis of the screw of the meter, and the observation would become useless. All trials in great depth on a definite level, or for mean current, should be made from a platform supported by two boats about 12 feet apart, in the centre of which there should be a well to allow the lowering and raising of the meter to the desired levels. This platform should be moored by four anchors to avoid oscillation in any direction.

All current observations require calm weather, without which they are of little or no value. On small rivers, a couple of hundred yards wide, it will not matter whether there be or not a strong breeze, because their surface is but little agitated by wind. Even with small rivers, however, calm weather is preferable and will ensure greater accuracy. With large rivers, exceeding say 500 yards in width, and with estuaries, no attempt at observation should be made in other than calm weather. A moderate breeze agitates their surface, making the platform unsteady, and more so a single boat for surface current trials. We may gauge a river in any weather; the result, however, will deserve the less confidence the more unsettled the weather may have been.

Taking the minute observation as the standard for comparison, an experienced observer may determine its commencement and termination within a quarter of a second, and accordingly obtain a result within about one-third per cent. near the truth. By extending the time of observation to five minutes, the result may be correct to within one-fifteenth per cent.; an accuracy rarely required. It is desirable, however, to depend on five-minute observations, not so much on account of the higher accuracy of the result in reference to time, but because there is always a slight oscillation of the boat to and fro, especially in very deep water and with a strong current; and, as these oscillations are periodical, they eliminate one another by extending the time of observation. Five minutes, moreover, is a convenient space of time, and should always be made in combination with the minute observation; neither is of much value without the other; the two combined and agreeing, give an observation which may be relied on.

A section of a river having been determined, and the preliminary field operations having been completed as described in the preceding pages, the currents at the section may be determined as to velocity and position in the following manner:—For large rivers and estuaries with a variable current, usually due to the effect of tides, a permanent observatory should be moored at a convenient point on the line of section in the deeper part of the river. The observatory is necessary for reference of increase or decrease of current during observations, extending over many hours of the day, and sometimes without intermission day and night. At the observatory hourly trials should be made, always noting the exact time of the trial, and from these we obtain the nature of the currents and the mode of their increase or decrease, readily represented by a diagram. The observatory is also convenient for trials of mean currents, and to determine currents at any depth below the surface. The position of the observatory, moored on the line of section, is at once fixed by taking the angle under which the base appears. To ascertain surface currents at various points on the line of section, the engineer should take a small boat with a couple of sailors, provided with two small anchors and several hundred yards of strong line. The boat is then rowed from 50 to 100 yards above the point at which the current is to be observed, an anchor dropped, and then allowed to drift down with the current and to pass the line of section about 50 yards, when the second anchor is also dropped; the boat may now be hauled up by the rope of the first anchor until on the line of section indicated by the flags, upon which the rope of the second anchor should be drawn tight, and the boat is ready for current observation, its position being fixed by the sextant in half a minute. The surface current is then ascertained by the meter in the mode above described, taking always at least two independent observations, and in case there should not be a close agreement between the minute and five-minute observations, several should be taken; and after completion of the trial the angle under which the base appears should again be determined as a check on the first, and to ascertain whether the position of the boat had not changed. The sextant discovers the movement of a few inches if the base be of proper length. The depth may also be taken if the current be gentle and the depth moderate; but we do not recommend to mix the two up; it is usually impracticable, because no reliable sounding can be taken from a fixed position in a strong current, and because it is important to complete the measurement of the surface currents across the whole section in as short a time as possible on account of a possible change in the currents. With rivers subject to the influence of tides, the currents across the section should be observed at high or at low water, because they will remain nearly the same for hours; all of which may be determined from the gauge at the margin of the river, and from the currents at the observatory. Our remarks apply to large rivers and estuaries; with small rivers, and a depth not exceeding 10 or 12 feet, the permanent observatory is not necessary; all the trials for surface and mean currents and for currents at a certain depth below surface, may be made from a small boat anchored fore and aft as above described for surface observations.

# CHAPTER III.

## The La Plata.

### THE ESTUARY.

THE La Plata, also called "River Plate," is a large estuary of the South Atlantic, from which the sea is constantly being displaced by the waters of the Paraná and the Uruguay. These two great rivers keep it filled with sweet water, and this circumstance may have originated the idea of calling the estuary a river, which in no manner it is. It has no drainage area of its own, nor has it the form of a river; it is a vast shallow basin, into which the Paraná and Uruguay pour their immense volume of water.

The La Plata is at its narrowest 23 miles wide, and is about 125 miles long; at its junction with the sea from Montevideo in a S.W. direction, its width is 63 miles, and here its water is already considerably mixed with the sea. For about 100 miles the water of the La Plata is sweet and pure, of a yellow tinge and opaque. The depth of the La Plata may be taken on the average at three fathoms or 18 feet at low water; it nowhere exceeds six fathoms or 36 feet, and from its narrowest part near Colonia, the depths slowly and uniformly decrease from about 21 feet as the estuary is ascended and the mouths of its feeders, the great rivers, are approached. Here considerable depressions or channels were formed by the currents, giving increased depth to navigation for some distance, when again shallows intervene and interfere. The material of the bottom of the La Plata is variable; in the shallower portions it is a very fine sand, impalpable and hard; in the deeper portions it is ooze of a neutral tint, soft and of a sticky nature.

In ancient time the estuary which now forms the La Plata reached about 200 miles higher up, and terminated in longitude 60° 35' west, latitude 32° 4' south, at a point now called Diamante, and from which the delta of the Paraná commences. The original length of the estuary, or of the ancient La Plata, was accordingly about 325 miles, of which 200 miles have been reclaimed by the Paraná; and in course of time, in the distant future, the whole of the La Plata will form part, and will be merged in the Paraná, the mouths of which will be where now the La Plata terminates, and it may then resemble the present delta at the mouth of the Mississippi.

The changes which take place in the La Plata are slow and certain, and are not capricious. Our observations have in a great measure determined the point. These have conclusively shown that with a given fall, whatever it may be, the current at the surface is proportional to the depth; accordingly, at double the depth we have double the surface current; at treble the depth the current will be three times as strong, everything else remaining the same; further, that the current at the bottom increases with the depth more rapidly than the surface current. If we hold this law in view—determined from observations on the La Plata and the Paraná—and consider its operation, the gradual formation of banks, of islands, and channels of great depth, is readily explained. Wherever the estuary is shallower the current is weaker, and deposit of earthy and fine sandy matter will not only take place more rapidly, but when deposited it is less likely to be removed by the current, which may be only one-half or one-third of that existing at the same time in the greater depths. The La Plata holds about $\frac{1}{16000}$th part by weight of solid matter in suspension. Deposit to any appreciable degree only takes place in still water, which in the La Plata, subject to tides, occurs twice a day and sometimes more often; and may last several hours, depending on the locality and its distance from the sea and the amount of tidal range. Deposit will take place equally over the whole surface within which for a time the currents are checked. With the turn of tide the current will re-establish itself according to the fall, and its strength at the bottom will be governed by the depth, and it will be stronger in the deeper than in the shallower portions of the estuary. The stronger current in the greater depths may disturb and remove deposit; the weaker current in the smaller depths may partly disturb and remove it; certain it is, it cannot do so to the same degree, and consequently the deposit in the shallower parts must take place more rapidly than in the deeper portions of the estuary; and, as the shallower parts get more and more shallow, so the deposit will be greater and greater, until it becomes shallow enough for the growth of rushes, and then the bank will grow rapidly; for, rushes and weeds check the current at all times and reduce it to a minimum. An island is now forming. The deeper portions of the estuary have, during all the time the island was forming on the shallows, changed but little, if changed at all. With the formation of the island, especially from the time the rushes made their appearance, the sectional area for the discharge of the volume of water is getting reduced, and on account of tidal range alone the inclination or fall must be increasing as the island is forming, because the free propagation of the tidal wave will be more and more checked, which will cause a greater difference of level between two points a certain distance apart. The increasing fall, due to tidal action, will increase the strength of the current in the deeper parts, but can have no sensible effect on the shallows already covered with rushes, and now the deeper portions of the estuary will be further deepened by the current, increasing in strength as the islands are forming right and left, which ultimately will be continuous banks, and the deeps will form the new channel of the extended river. By this process the La Plata will merge into the Paraná at some distant future time, and the Palmas and Guazú branches will extend beyond Montevideo, and the La Plata will be no more, islands and lagoons having taken its place.

## TIDES.

In all new countries, observation, the foundation of sciences, is but little known and less practised. That the La Plata was subject to tides was not known by thousands of people living for generations on its very margin. The existence of a tide was not admitted by those supposed to be best informed on the subject, and we were informed that there was no tide, and that the rise and fall of the surface of the La Plata was due to the effect of wind. We have thereon established systematic observations, recording every quarter of an hour the position of the surface of the La Plata by a gauge; the force and direction of the wind; the position of the barometer and thermometer; and these observations were continued day and night without intermission during the whole time of our surveys. The question whether there was or not a tide in the La Plata was soon determined, and one month settled the point which a number of succeeding ones confirmed. It could only have been a question how far the tide of the South Atlantic was propagated, and not whether there was or not a tide; for, the La Plata and the sea are in a great measure one and the same thing. When we speak of a tide, it is understood to be regular and lunar—a periodical rise and fall of the surface of the water produced by the attraction of the moon and sun, especially by the former; and that the position of the surface of the water will chiefly depend on the moon in reference to the meridian of the locality under consideration. The tidal wave of the South Atlantic enters the La Plata estuary and propagates itself on its whole extent with almost unbroken force. It continues its progress and enters the mouths of the Paraná and the Uruguay, and travels up these rivers a hundred miles and more, until it becomes so much reduced that it seems to disappear, and escapes observation.

On the day of new or full moon there is high water at Buenos Ayres at 6 P.M. Low water corresponds nearly with the meridian passage of the moon. Generally speaking, whenever the moon is on the horizon, it is high water at Buenos Ayres; whenever she passes the meridian, the upper or the lower, it is low water. The transit of the moon is every day 49 minutes later on the average; and the observations show that the tidal wave is also 49 minutes later on the average. The meridian passage of the moon is, according to her position in her orbit, from 42 to 62 minutes later every succeeding day; and observations show that the tide at Buenos Ayres faithfully follows the variation in time of the moon's meridian passage. Every 24 hours and 49 minutes, which constitute a lunar day, there are two tides, corresponding to the meridian passage of the moon above and below the horizon; the time of high water happens always a certain number of hours after her meridian passage, and we have two waves corresponding to her upper and lower transit. If the moon be in the equinox the two waves are of equal height; as her declination is getting more and more south, the wave corresponding to the upper transit of the meridian is getting larger, and that corresponding to the lower transit smaller, until the maximum declination of the moon is reached, and then we have a high wave corresponding to the upper, and a small one nearly merged in the large wave, corresponding to the lower transit. As the moon again approaches the equinox, so the two waves get more and more alike; and when she enters the northern hemisphere, the first wave corresponding to the upper transit becomes the smaller and the second the larger, and all the succeeding phenomena are similar except reversed. These observations only show what happens at every place where tides are carefully observed; and to deny the existence of a tide on the La Plata is nothing short of denying the existence of the moon. The influence of the sun may also be traced by his bringing the tide on a little sooner or later, as the moon appears in her first or last quarter.

The influence of wind on the surface of the La Plata is great, and depends on its strength and the quarter from which it blows. A strong wind from the east or south-east raises the level of the La Plata; the same wind from west or south-west depresses it; and from north or south does not sensibly affect its level. The extent to which the surface of the La Plata may be raised or depressed further depends on the duration of the wind;

but whatever may have been its effect in raising or lowering the surface of the La Plata, the tidal phenomena nevertheless take place with unerring certainty at that higher or lower level. There is, indeed, an effect of the wind on the tide in an indirect manner; that is to say, a wind having raised the level of the La Plata many feet, deepens the estuary all over, and the tidal wave travelling faster in the deeper water, will bring high water a little sooner on, perhaps an hour; whilst a wind which depresses the level of the La Plata will make the estuary shallower all over, and the free wave can but travel slower in the smaller depth; it will take therefore a little longer before the wave can reach Buenos Ayres from the South Atlantic; accordingly, low water will be somewhat later; and, cause and effect being clearly traced, there need be no further mystery about it.

On the diagrams of Plates II. and III. we have given the tide records for several days; Diagram No. 2, Plate II., shows the tide when the moon had just passed her first quarter, December 29, 1870, then with a southern declination of about 2 degrees. Here the difference between the two waves following the upper and lower transit is considerable. On the 30th December, 1870, the first wave W, due to the upper transit, is high; the second X, following the lower transit, is small and nearly absorbed by the larger one. The interference of the sun is here also greatest and most marked. For the upper transit high water was about an hour behind time; for the lower, about half an hour before time. This is the period of greatest disturbance, the moon being in her greatest change of declination and also at her greatest variation with the tidal action of the sun; the moon depressing the tides whilst the sun raises them, being the period of neap tides. The weather was favourable for observations, the larger part of the day and night being calm, and the breeze noted on the diagram was from a quarter which had no sensible effect on the surface of the La Plata.

During the afternoon of December 30th, 1870, the weather was very favourable for observations, and the tidal wave shown on the diagram represents the outline it assumes when nothing disturbs its regular movement. With ordinary low water at Buenos Ayres, our gauge indicated 3 feet 4 inches; that is, the La Plata always fell to that level when nothing disturbed the tidal wave. On the 30th December, the first high water due to the upper transit of the moon marked 8 feet 7 inches on the gauge; the top of this wave was therefore 5 feet 3 inches above ordinary low water. The second tidal wave, due to the lower transit of the moon which happened at 6h. 35m. 30s. in the morning, marked high water at 6 feet 5 inches; the top of this wave was therefore 3 feet 1 inch above ordinary low water. The difference in the height of the two waves is 2 feet 2 inches, from which it appears that on account of the rapidly changing declination of the moon the first wave was raised 13 inches above, the second was as much depressed below the normal height of wave, which for this day would have been 4 feet 2 inches had the moon not changed her position. It is of rare occurrence that a tidal wave may be observed free from disturbing causes. On the next day, viz. the 31st December, we had a variety of breezes from several quarters, nearly all of which had their effect in temporarily raising or depressing the La Plata, although the disturbance was on the whole small. This diagram will illustrate the outline of neap tides on the La Plata.

With Diagram No. 1, Plate III., we have an illustration of another series of tidal waves, as they appear at new moon or at spring tides; the moon passing the ecliptic the 19th January, 1871, in the afternoon. We have, moreover, a series of violent disturb-

E

ances by gales from different quarters, lasting for a considerable time. The diagram is made to represent another set of observations referring to currents and the propagation of the tidal wave; for the present we need only note the outline of the waves of the La Plata, marked A, B, C, D, E, F, and the meridian passages of the moon, marked at the hour of the day at which the transit took place. The disturbing causes, originating with astronomical circumstances, *viz.* the position of the sun and the moon, only partly exist in this diagram; the sun and moon being close together, act in the same sense and direction in producing the Atlantic tidal wave; and the moon, being near the ecliptic, and in her greatest declination 23° south, changes but little her position; all of which was just the reverse in the preceding diagram we have considered, representing neap tides of the La Plata. Looking at this diagram, it will be seen that the nearer we approach new moon, which happens on the 19th in the afternoon, the nearer coincides the upper and lower transit of the moon with low water, although the gale of the 19th and 20th raised the low-water line of the La Plata exceptionally high, being at 7 feet 6 inches instead of 3 feet 4 inches under ordinary circumstances; further, that the tidal waves of the 19th and 20th are similar for the upper and lower transit, although at very different levels, all which clearly shows that the tidal phenomena took place with unerring regularity and certainty, and that the wind only modifies and qualifies these phenomena, but does not in any way originate them. There are six tidal waves represented on the diagram. The first, marked A, was formed during a perfect calm; but before its termination at 8 P.M. on the 17th, a gentle breeze from the east sprang up, which prevented the low water falling to its normal level; the La Plata would have continued to fall till about 10 P.M., but the breeze raised the level more than it would have fallen on account of tide alone; in other words, the breeze made the tide turn sooner; and it should be noted how sensitive the La Plata is to winds from the east or near that point. The breeze from the east freshened to a gale of wind before midnight, and continued for nearly six hours without intermission with the same force, except that it veered to N.E. The effect of this gale on the tidal wave, due at about four o'clock in the morning of the 18th for high water, is remarkable; for, not only was this wave raised to exactly double the height of the preceding one, to which it would have been similar had the wind not interfered, but it made high water about two hours sooner, keeping the top of the wave three hours stationary. At five in the morning the gale ceased, blowing a strong breeze, but at the same time it veered round to the north, and now the opposition of the wind to the fall of the level of the La Plata being withdrawn, its level falls rapidly from a height which was double that belonging to the tidal wave alone; and so great was the momentum of the immense mass which descended from the higher level, that the fall continued two hours beyond the time at which low water would be reached by the tidal wave alone, independently of the wind. At about noon the wind ceased, and there was calm weather for many hours, and we have again the regular formation of a tidal wave with high and low water corresponding to the meridian passage of the moon, *viz.* low water at the time of the meridian passages, high water about six hours later. The calm at Buenos Ayres practically continued for about twenty-four hours, a gentle breeze occurring from various quarters, all of which had little or no effect on the surface of the La Plata, chiefly because these were shifting and from opposing quarters.

It may at first sight appear strange how the wave D, under nearly similar circumstances of a gentle shifting breeze, and from a quarter which has little or no effect on the La Plata, should have been so much the greater than the one preceding it; there is apparently nothing to show on the diagram that wave D should have been larger than wave C—certainly nothing to make it nearly treble the height. There is, however, a cause for everything in nature we observe, and if we should not see it at once, we must inquire, consider, and seek to find it out. A storm on the Atlantic may travel at a speed of say 30 miles an hour; a free wave, raised by the moon or any other cause, would in the great depths of the Atlantic travel at about 300 miles an hour. Suppose a storm was raging east of Buenos Ayres at 2 P.M., January 18th, 1871, at a distance of 840 miles, approaching Buenos Ayres at a rate of 30 miles an hour, it would reach the town in twenty-eight hours, and appear at 6 P.M., January 19th. The tidal wave, however, which at 2 P.M., January 18th, was also 840 miles distant, and which was exposed to that storm and had its height increased by the force of the wind acting in the same direction, will reach the margin of the La Plata in little under three hours; and the same wave in the shallows of the estuary would take about six hours more to traverse it as far as Buenos Ayres; accordingly, the tidal wave would reach from the locality at which the storm was raging at 2 P.M., January 18th, to Buenos Ayres in about nine hours, and would make its appearance at 11 P.M. the same day. This is what happened with wave D. It came in advance of the storm, which increased its height; but when the top of the wave was gauged at Buenos Ayres there was no sign of a storm, indeed there was at the time calm weather with gentle currents of air at 90 degrees with the approaching gale. All this may happen with storms approaching from the sea, namely, from the E., S.E., and N.E., and especially with storms from the S.E., from which quarter the tidal wave approaches. Storms from other quarters can have only a local effect on the La Plata itself.

At the termination of wave D at 11 A.M., January 19th, and at the commencement of the next wave E, the advanced part of the storm had already arrived at Buenos Ayres in the shape of N.E. breeze, which, increasing in strength rapidly, was within two hours a strong breeze from the east; all this time the wave E was passing along the city, and during high water the storm reached Buenos Ayres with its full strength at 6 P.M., January 19th. Both waves of the 19th January were exposed to the effects of the same storm, and they are of identical height, reckoning each from the preceding low water, and the only difference is, that by the time wave E had arrived the surface of the La Plata was by the approaching storm already raised 22 inches above ordinary low water, and that after the storm had spread over the estuary and for many hours continued to press it westward, moving large masses from the Atlantic with it, the surface of the La Plata, independently of the tidal wave, was raised an additional 28 inches; so that low water after the termination of wave E was 50 inches above ordinary low level, an extraordinary event,—which produced a kind of revolution on the territory of these great rivers, and for a hundred miles up the stream reversed the current of the mighty Paraná.

The next tidal wave F reached the La Plata at an exceptionally high level, and although there was still a strong breeze at Buenos Ayres while it arrived and formed the next high water at 4 A.M., January 20th, the F wave had no storm in the Atlantic, and it was only as it entered the La Plata that a strong breeze began to favour its movement, which soon moderated to an ordinary breeze, releasing the waters of the La

E 2

Plata which were pent up at an exceptional level by the storm which had just passed by. The wave F is but an ordinary tidal wave at an exceptionally high level, and although its top is the highest above ordinary low water of the various waves we have just considered in succession, it did not by any means produce the revolution on the great rivers which followed the formation of the E wave.

Tidal phenomena are most interesting and instructive to hydraulic engineers, and we have dwelt at some length on the tides of the La Plata, because, what has been observed and said in reference to the La Plata holds good with every estuary on the Atlantic, and it will only be a question of degree depending on local circumstances. The tides we have just considered are not selected from a number of tidal observations, but taken for consideration because it happens, that on those days important observations on currents had been made on the La Plata and the Paraná de las Palmas, which can only be understood and explained by a careful consideration of the tidal waves and their effect and interference with the currents of the estuary and the great rivers. Nor has the subject of tides on the La Plata been fully considered or exhausted by the consideration of some neap and spring tides; and although the causes producing any other tidal wave will be similar or differ but little from those we have investigated, the phenomena present themselves in ever-changing variety, and it would probably be difficult to find two waves of identical formation in observations spreading over a whole year embracing about seven hundred tides.

## CURRENTS.

The La Plata being an estuary of the Atlantic subject to tides, its currents are ever changing in velocity and also direction. A further complication in regard to the currents of the estuary arises from the circumstance that the La Plata is at the same time receiving an immense volume of water from the great rivers, which it has to convey to the Atlantic. Although the volume of water discharged by the great rivers into the La Plata is for months together sensibly a constant quantity as far as the rivers are concerned, the volume actually discharged per minute by the rivers at their mouths and finding its way into the estuary, is ever varying and constantly changing, on account of the tidal rise and fall of its surface. The tide alone would produce a diversity of currents depending on the position of the tidal wave, namely, how far it had advanced on the La Plata. The tidal wave having entered the mouth of the estuary at one end, the discharge of the great rivers at the other end will be modified; and these circumstances bring about such complications involving areas, volumes, depths, inclinations, and time, that nothing short of actual observation at the locality under consideration can determine the current we wish to ascertain. In some reaches of the estuary it would be difficult to say whether there was or not a current at a given time, and we could not tell beforehand whether its direction would be towards the Atlantic or the reverse; observation alone can determine the point.

The currents of the La Plata are by no means similar to those of a river; we have here simultaneously currents in all directions and of every variety. This is especially the case with the approach of a tidal wave, that is, during the rise of the surface of the estuary. When the wave is passing away, that is, during the time from flood to ebb, the currents have greater similarity in direction and velocity. As the tidal wave enters the mouth of the La Plata, a current is soon established into the estuary from the Atlantic.

At the same time there is a strong current from the great rivers into the estuary at its other end, the La Plata filling from opposite quarters and in opposite directions.

As the wave advances it proceeds faster in the deeper portions or channels, and in these it is raising the level quicker than in the shallower parts near shore, where the same wave will raise the level to the same height a little later, its progress being retarded by the smaller depth of water; and, as water always follows inclination, a current arises from the deeper channels of the estuary towards the shore; in other words, the set of the current is now on to shore. A variety of currents arise on account of the different depths within which the tidal wave proceeds, and the different volumes of water required to fill succeeding parts or sections of the estuary of varying widths. As the wave proceeds higher up, first checking and then reversing the current which previously was to the sea, the volume of the great rivers begins to tell on the wave, the progress of which they cannot arrest, but a contention arises as to checking and reversing the current, and the nearer the wave approaches the rivers the more powerful they become in maintaining their established current, and the less can the tidal wave check or reverse them.

It now depends on the height of the wave and on the level on which it approaches and enters the estuary, at which point its power will balance that of the rivers; power being measured by the volumes of water which either of them can bring to bear per minute in raising the level of the La Plata. At this point, where the power of the wave balances that of the rivers, there will be no current; the level of the estuary will rise slowly like that of a lake receiving supply from all round its border. It is here—where the rivers and the tidal wave contend for supremacy, each trying to establish its own current, and where for hours the power of either of them trembles in the balance without any sensible movement in any direction—that deposit copiously takes place; matter, held in suspension by the rivers as long as their currents are maintained agitating their water, is dropped as soon as they come to rest. It is here, within about 10 to 20 miles of the rivers' mouths that banks are most rapidly growing and islands are forming, and the ultimate result of these daily contests is invariably in favour of the rivers which slowly but steadily encroach on the estuary and ultimately annex its whole territory. The progress of the tidal wave is, however, never checked an instant, the rivers only check the currents originating with the wave; and from the point where the power of the rivers and the wave balances, it is by the current of the rivers that the tidal wave propagates itself, ascending them, and they supply the volume necessary for its formation; and, as by the inclination of the rivers' surface the top of the wave approaches more and more their surface, it becomes smaller and smaller, and ultimately imperceptible; and it dies out as it ascends the rivers within about 100 to 150 miles from their mouths. A tidal wave is never visible to the eye, and can only be conceived from observation, by a successive measurement of its dimensions which are very large. We may, from an elevated position, see 10 or 15 miles, but a tidal wave on the La Plata is about 258 miles long; the waves we do see are local, and due to the agitation of the surface of the water by the wind, and are only as many ripples on the invisible tidal wave.

The preceding considerations of currents refer to a period when the La Plata rises from low to high water, that is from ebb to flood tide. In the lower reaches of the estuary near the Atlantic, as soon as the top of the wave has passed a certain locality, the currents which for many hours were into the estuary are by degrees checked and reversed, now setting

from the estuary into the Atlantic. Higher the currents are sooner reversed towards the
Atlantic, until a locality be reached where the power of the rivers and the wave balances,
and here the current is no more reversed at any time but reduced to nil; still higher up the
current is only reduced in velocity, but its direction remains the same at all time, viz. from
the rivers to the Atlantic. During the second half of the tidal wave, viz. from flood to ebb
when the surface of the La Plata is falling, there is much more uniformity in the direction
of the currents, which for a time will be the same for the whole estuary, all tending to the
Atlantic. The wave will again proceed faster in the deeper than in the shallower portions
of the estuary, and will accordingly make the level fall a little faster in the deeper channels,
and the current will now set from shore into the estuary; the reverse of what happened
with the rise of the La Plata.

By degrees the level of the estuary will again adjust itself to mean sea-level. All the
water which the tidal wave brought from the sea will now have to be returned, and in
addition the whole volume which the great rivers have discharged into the estuary; and the
currents will not only be stronger but they will also last longer, of which circumstance the
outline of the tidal wave bears evidence; the duration of the rise of the La Plata being
about six hours, its fall continuing for about seven hours. These general observations
referring to the currents of the La Plata were collected while we were engaged with
various surveys along its margin and while we navigated the estuary in various directions
for days and nights together.

## CURRENT OBSERVATIONS.

Systematic current observations as to their definite velocity and direction were made
about three miles off Buenos Ayres in a gentle depression of the La Plata, affording good
anchorage for large vessels, and called the "Outer Roads." From time to time a number of
trials were made, and we select a set of consecutive observations favoured by exceptionally
fine weather, which are the more interesting inasmuch as there is no gap in the records,
which happens only too often through accidents or other causes not within the power
or control of the engineer.

On the morning of the 30th December, 1870, we sailed with the steamer at our disposal to the Outer Roads, and after fixing the position of a prominent wreck close by, a
section was made across the direction of the current, to obtain apart from charts an exact
knowledge of the locality. To illustrate the whole proceedings we will give the details of
the survey with the figures of the observations, premising, that our fixed points on shore
were the steeples and domes of the city of Buenos Ayres, their exact position having
previously been determined by triangulation; and any locality on the estuary, as long as
these points were visible, was determined by the angle under which at least three of them
appeared, and which was readily done by the sextant.

The soundings were taken from the wreck in the Outer Roads on a line with the
Catalina Church, the steamer proceeding at a very slow pace, and each sounding was
usually one minute in time apart from the other; at the last sounding the anchor was
dropped and the position of the steamer fixed by triangulation; we had accordingly the

length of the distance traversed and a number of nearly equidistant soundings on the line between the wreck and the last point.

The following are the entries in the Survey Book :—

DECEMBER 30TH, 1870. LOCALITY—WRECK IN THE OUTER ROADS, LEADING MARK OF CATALINA CHANNEL.

Angle observed by
large Sextant.

| No. 1. | 5° 28' 20" | between Churches Telmo and Francisco. |
| „ 2. | 8° 7' 50" | „ „ „ „ Cathedral. |
| „ 3. | 28° 9' 50" | „ „ „ „ Recoleta. |
| „ 4. | 13° 19' 0" | „ „ „ „ Catalina. |

Magnetic bearing from Wreck to San Francisco, 215° 30'.

Any two of the above angles will determine the position of the wreck in reference to the churches of the city, their steeples or domes, and may be laid down on a plan embracing the city and the shore of the La Plata ; the others are check angles ; the position is readily determined according to well-known principles of nautical surveys. The angles which will give the position with the greatest accuracy are Nos. 2 and 3, and No. 4 may be taken as a check on the result. The triangles Telmo, Cathedral, **Recoleta** ; and Telmo, Cathedral, Catalina ; being known from the triangulation of **the** city, or any other combination of triangles among these points, the position of the wreck in reference to those points is readily found either by calculation or by construction, and with large angles offering favourable intersections, the latter mode of plotting the lines and angles will be the quickest and most convenient for ordinary purposes. We almost invariably followed it, and the check intersection usually came within the thickness of a pencil line from angles determined by the **pocket sextant.**

The soundings taken on the line, Wreck—Catalina, commencing at the wreck, and reduced to ordinary low water of the La Plata marking on our gauge 3 ft. 4 in., were 25 in number, with the following depths taken from the diary :—

" DECEMBER 30TH, 1870. LOCALITY—OUTER ROADS."

| No. | Time. | Depth. | Remarks. | No. | Time. | Depth. | Remarks. |
|---|---|---|---|---|---|---|---|
| | P.M. | ft. in. | | | P.M. | ft. in. | |
| 1 | 1 38 0 | 15 4 | At wreck. | 14 | 1 51 15 | 17 4 | |
| 2 | 1 39 0 | 18 6 | | 15 | 1 52 30 | 14 6 | |
| 3 | 1 40 0 | 21 4 | | 16 | 1 53 0 | 13 4 | |
| 4 | 1 41 0 | 21 4 | | 17 | 1 53 30 | 13 4 | Soundings re- |
| 5 | 1 42 0 | 22 6 | Soundings re- | 18 | 1 54 0 | 13 4 | duced to ordinary |
| 6 | 1 43 0 | 22 6 | duced to ordi- | 19 | 1 55 0 | 13 2 | low-water level, |
| 7 | 1 44 0 | 22 6 | nary low-water | 20 | 1 56 0 | 12 6 | 3 ft. 4 in. on |
| 8 | 1 45 0 | 21 4 | level, 3 ft. 4 in. | 21 | 1 57 0 | 11 4 | gauge. |
| 9 | 1 46 0 | 20 6 | on gauge. | 22 | 1 58 0 | 11 6 | |
| 10 | 1 47 0 | 18 4 | | 23 | 1 59 0 | 11 8 | |
| 11 | 1 48 0 | 18 4 | | 24 | 2 0 0 | 11 8 | |
| 12 | 1 49 0 | 18 4 | | 25 | 2 1 0 | 11 4 | Dropped anchor. |
| 13 | 1 50 0 | 16 6 | | | | | |

Position of last Sounding, No. 25.

Angle observed,

| No. 1. | 9° 43' | between Churches Telmo and Cathedral. |
| „ 2. | 33° 58' | „ „ „ „ Recoleta. |
| „ 3. | 15° 6' | „ **Telmo and Tank** of Waterworks. |

The angles at the last sounding determine its position, and accordingly the distance from the wreck is known. The time between the first and last sounding is 23 minutes, and if we divide the distance into 23 parts, we have the position of soundings for the full minutes, and the intermediate ones are also fixed by the fraction of minutes. A section of the locality for current observations having been obtained, a platform was erected near sounding No. 6 on the line of section, as already described in the preceding chapter, pp. 19 to 21, and all being prepared, current observations were made with the whole apparatus for the integration of currents and their measurement at definite levels.

We shall again give the entries as noted in the Survey Book, from which afterwards the diagrams were prepared; the proceedings are more readily followed by giving the notations of the observations themselves, explaining afterwards how these had been worked out for the diagrams.

"DECEMBER 30TH, 1870, AFTERNOON. LOCALITY—OUTER ROADS."

Slight breeze, surface of water a little undulating. Meter No. 1 attached to Integrator. Observations made from platform resting on two boats 12 feet apart.

| No. of the Observation. | Time of Commencement and Termination. | Time of Observation. | Indexes of Meter. | Position of Meter. | Remarks. |
|---|---|---|---|---|---|
| | h. m. s. | m. s. | | feet. | |
| No. 1 | 3 30 0 P.M. / 3 36 2 „ | 6 2 | 0a + 0 / 2a + 120 | 4 | Below surface. |
| „ 2 | 3 38 0 „ / 3 41 5 „ | 3 5 | 2a + 120 / 3a + 185 | 4 | „ „ |
| „ 3 | 3 43 0 „ / 3 46 4 „ | 3 4 | 3a + 185 / 5a + 23 | 7 | „ „ |
| „ 4 | 3 48 0 „ / 3 50 58 „ | 2 58 | 5a + 23 / 6a + 81 | 4 | „ „ |
| „ 5 | 3 53 30 „ / 3 56 30 „ | 3 0 | 6a + 81 / 7a + 90 | 10 | „ „ |
| „ 6 | 3 58 45 „ / 4 1 45 „ | 3 0 | 7a + 90 / 8a + 155 | 4 | „ „ |
| „ 7 | 4 4 0 „ / 4 7 3 „ | 3 3 | 8a + 155 / 9a + 111 | 16 | „ „ |
| „ 8 | 4 9 30 „ / 4 12 33 „ | 3 3 | 9a + 111 / 10a + 189 | 4 | „ „ |
| „ 9 | 4 20 0 „ / 4 21 0 „ | 1 0 | 0a + 9 / 0x + 38 | 1 | From bottom. From this observation indexes set to zero each time. |
| „ 10 | 4 25 0 „ / 4 27 0 „ | 2 0 | 0a + 63 | 1 | From bottom. |
| „ 11 | 4 30 0 „ / 4 32 0 „ | 2 0 | .. | 1 | From bottom. Meter not operated. |
| „ 12 | 4 37 0 „ / 4 38 0 „ | 1 0 | 0a + 47 | 1 | From bottom. |
| „ 13 | 4 42 0 „ / 4 43 0 „ | 1 0 | 0a + 38 | 1 | „ „ |

| No. of the Observation. | Time of Commencement and Termination. | Time of Observation. | Indexes of Meter. | Position of Meter. | Remarks. |
|---|---|---|---|---|---|
| | h. m. s. | m. s. | | feet. | |
| No. 14 | 4 48 0 P.M. <br> 4 50 0 „ | 2 0 | 0a + 69 | 1 | From bottom. |
| „ 15 | 4 54 0 „ <br> 4 55 0 „ | 1 0 | 0a + 98 | 4 | Below surface. |
| „ 16 | 4 56 0 „ <br> 4 57 0 „ | 1 0 | 0a + 99 | 4 | „ „ |
| „ 17 | 4 58 0 „ <br> 4 59 0 „ | 1 0 | 0a + 87 | | Lost; level uncertain, either 7 or 10 feet. |
| „ 18 | 5 0 0 „ <br> 5 1 0 „ | 1 0 | 0a + 82 | 10 | Below surface. |
| „ 19 | 5 6 0 „ <br> 5 11 0 „ | 5 0 | 0a + 0 <br> 1a + 181 | Surface to bottom. | Three times down and three times up in succession, for mean current. |
| „ 20 | 5 14 0 „ <br> 5 19 3 „ | 5 3 | 1a + 181 <br> 4a + 71 | 4 | Below surface. |

" Depth of water under platform, 25 ft. at 2.25 P.M."

" A storm from shore was approaching rapidly at 5.30 P.M., with lightning and thunder, and we had to clear off as soon as possible."

" Sample of water from near the bottom of the Outer Roads taken, also sample of mud from surface of bottom."

" At 6.45 P.M., current still running towards the sea—direction of current 125° by prismatic compass."

" At noon the current towards the sea hardly perceptible."

" All the afternoon a gentle breeze from the north, practically calm weather, surface but little agitated."

" At 7 P.M. the tide still running to sea with moderate speed. The 'Aguila' bearing anchor. Storm over Buenos Ayres. Perfect calm in the Outer Roads. Vivid lightning."

" Distance of platform on which current observations were made from wreck, measured by Meter No. 1, equal to 1136 Revolutions, viz. $\left\{\begin{array}{c} 4a + 71 \\ 10a + 1 \end{array}\right\}$ position of indexes."

Before we proceed to work out the observations, we shall explain the notation adopted for the meter, and the circumstances attending some of the trials and the mode we followed with these current measurements.

Under the heading "Time of Commencement and Termination," the exact time of the day is noted at which the instrument commenced and terminated to register the revolutions of its screw. It is not enough to know how many minutes and seconds the meter was exposed to the current, but the time of the day at which that current existed must be also known. The current differs and changes all the day more or less, according to the position of the tidal wave registered by the gauge on shore; and it must be referred to and considered with the gauge observations. The next column, "Time of Observation," gives the time in minutes and seconds during which the meter registered the revolutions of the screw. This is simply the difference of time between the commencement and the termination of the observation. Under the next heading, " Indexes of Meter," the position of the index on each worm-wheel is noted. One wheel bears the letter $a$, to distinguish it from the other wheel and to avoid confusion. The index of one wheel registers one division for one revolution of the screw. The index of the other wheel, marked $a$, registers one division for 201 revolutions of the screw. If for example we take No. 1 observation, we find in this column first " 0a + 0," which means, the position of the index was 0 on the $a$ wheel

F

and also 0 on the other, and that both readings added together make up the number of revolutions of the screw; the indexes standing here both at zero at the commencement of the trial. At the termination of the observation we find "$2a + 120$," which means, two divisions of the $a$ wheel and 120 of the other, being equal to $2 \times 201 + 120 = 402 + 120 = 522$ revolutions. The former position of the indexes being zero, it follows that during the time of the trial, given by the preceding column at 6 minutes and 2 seconds, the screw of the meter registered 522 revolutions. So, during the next trial, No. 2, the position of the indexes at the commencement is noted at "$2a + 120$," which also shows that the instrument was not used or altered since the termination of the first trial preceding it; the second reading of indexes being $3a + 186$, equal to $3 \times 201 + 186 = 789$ revolutions, of which 522 belong to the former trial, the difference between the two readings of the indexes at the commencement and termination will be the number of revolutions due to the second trial, viz. $789 - 522 = 267$, performed in 3 minutes 0 seconds of time; and so on every following notation of this column will be readily understood. The next heading, "Position of Meter," explains itself; it registers the position in which the meter was held during the trial, viz. 4 feet below surface, or 7 or 10 feet, as the case may be.

At the commencement of trial No. 9 the indexes were both set to zero for each wheel, and so with every succeeding trial until No. 20. These were short observations not exceeding two minutes, chiefly minute observations; and the intention was to make each succeeding reading and observation independent of the preceding one, especially since by the nature of the current and the shorter time of the trials one wheel registered all the revolutions. There is the advantage in this proceeding, that an error in the reading of the indexes is not transferred to the following one, which would therefore vitiate only one trial and not two. On the other hand, it has the disadvantage of constantly adjusting the instrument to zero, and the continuity of the records is lost. It depends on circumstances which proceeding should be preferred. If the instruments are so constructed that the reading of the indexes is clear and rapid, involving no doubt and loss of time by the use of magnifying glasses, it is by far the best plan not to change their position but to continue the readings; the only precaution needed is to read the indexes twice —viz. at the commencement and termination, and not to copy from the Survey Book the position of the indexes as noted for the termination of the preceding trial as that for the commencement of the next observation.

Looking at the notations of these trials, it may appear that the principle laid down in the preceding chapter on current observations, page 18, was not strictly followed by having always two for the same current, the one as a check on the other. In the first place, it should be remembered, that the object of the check observation is not to check the current, but to check the accuracy and the correctness of the observation, which principally depends on the instrument being in perfect adjustment; and to ascertain this is nearly the whole object of the check observation; for, everything else may be checked by other means, by double readings and double entries, &c., &c. Moreover, in estuaries or rivers subject to a tidal rise and fall, the current is at times changing, so that observations only five minutes apart, if exact, may register different currents; and the check observation, which cannot follow much quicker on account of depth and necessary arrangements, may differ from the preceding one considerably, and we are at the time left in doubt whether this difference is due to the meter or to the current. Under such circumstances it is preferable to make the check observations at a fixed level, which may serve as a datum for the whole series of trials.

In the above series, the datum for comparison of relative currents and as a check on the instrument was a level 4 feet below surface, on which the check observation was made after each trial at levels much below the surface. This level was taken in preference to the surface in order to have the full current operating on the screw, which at the surface might have been a little reduced by the proximity of the boats supporting the platform, about 6 feet from either boat to the centre of the meter. So, after the 7, 10, and 16 foot levels, the check observation for the meter was invariably made at the 4-foot level. The observations from No. 9 to No. 14, both inclusive, were exceptional, and all on the same level, viz. 1 foot from the bottom of the La Plata. We were anxious to ascertain the current near the bottom. No. 9 was the first observation. No. 10 followed as promptly as possible; and the result differing 6½ revolutions, we did not consider it a satisfactory check, and made another, No. 11. This was somehow lost, either the wire was not drawn tight enough, and did not throw the meter into gear, or, what is more likely, the weights sunk into the soft mud, stopping the movement of the screw while thrown into gear. Then No. 12 followed, the result of which differed nine revolutions from the first observation, which again was not considered a good agreement; then No. 13 and No. 14, which came nearer, but the oscillations of the records were much wider than usual, even with currents of treble the velocity; and, beginning to doubt the condition of the meter, we immediately made two observations in succession on the 4-foot datum level as a check on the meter, which, however, only proved by their close agreement, that the instrument was correct and in perfect adjustment, and that the records of the meter near the bottom probably truly represent oscillations of the current itself. The diagrams illustrate this circumstance strongly, though they do not finally determine the point; we shall more fully refer to it when the diagrams themselves will be under consideration. Observation No. 17 was lost, because a doubt arose whether the trial was on the 7 or the 10 foot level, and in experiments of this nature nothing should be assumed, and positive information alone trusted. No. 19 is an observation for mean current by using the apparatus as a current integrator, and is very accurate and reliable; also shown by the next trial at the datum level, the instrument recording a current within four-tenths of a revolution the same as No. 15, about 25 minutes previously; showing that the current was practically constant and on the turn, having reached and passed its maximum; it also closely agrees with the tide records near shore, giving the maximum current at half tide; from flood to ebb; the greatest inclination of the tidal wave was at that time.

The above observations determine the current in feet per minute for the precise time of the day by substituting for the revolutions their value in feet, as ascertained by experiment. We have already mentioned, page 15, that the value of one revolution of the screw is a variable quantity, and will depend on the velocity of the current. The value of each revolution can therefore only be ascertained by trial in moving the meter in still water at different velocities, and by recording the number of revolutions for the same distance. If the resistance of the mechanism of the meter were nil, which would indicate the absence of all friction whatever, the revolutions of the screw would be proportional to

F 2

the velocity of the current, from nil increasing to any extent, and the increase in the number of revolutions due to the increase in the velocity of current would be represented by a straight line, starting from the intersection of two co-ordinate lines at right angles, in which the horizontal one, or the abscissa, may represent the velocity of current in feet, and the vertical, or ordinate, the number of revolutions of the screw per minute. Calling the abscissa of any point of the line $x$ and the ordinate $y$, the equation of the line would be $y = a\,x$, where $a$ represents a constant quantity, viz. the tangent of the inclination of the line, which will depend on the pitch of the screw. There is with every instrument, however delicate, a certain amount of friction which requires a certain current to overcome, and until that current be reached the instrument will not turn, and register zero. After that current be reached, with every further increase the revolutions of the instrument will be sensibly proportional to the increase of the current, as shown by direct experiment. The line representing revolutions will therefore be very near a straight line, not originating however from the intersection of the co-ordinates, but intersecting the abscissæ some distance from that point. With current meter No. 1, used during our observations, the general equation of a line, viz. $y = a\,x + b$, was for the pitch of the screw and the friction of the instrument determined by experiment found to be $y = 0\cdot9962\,x - 6\cdot811$, where $y$ represents the number of revolutions of the screw per minute, and $x$ the velocity of the current per minute in feet; and since we always obtain the revolutions by an observation, and want to find the current ($x$) which produced them, we have from the above equation of meter No. 1, $x = 1\cdot0038\,y + 6\cdot837 =$ current in feet per minute, having to substitute for $y$ in the preceding equation the number of revolutions observed per minute. We have gone more fully into this subject in the Appendix on " The improved Current Meter."

Applying the equation of the meter for the preceding observations, we obtain in each case the current in feet per minute; for example, in observation No. 1 we have $(2a + 120) = 522$ revolutions, in 6 minutes 2 seconds, or per minute $86\cdot52$ revolutions, which would be the value of $y$ for this observation. Substituting this value in the equation we have $x = 1\cdot0038 \times 86\cdot52 + 6\cdot837 = 93\cdot68$ feet per minute velocity of current as the result of observation No. 1; and so on with every succeeding trial, the currents are obtained with great accuracy.

The results of the observations of the 30th December on the currents of the La Plata appear from the following Table :—

TABLE I.

DECEMBER 30TH, 1870.   OUTER ROADS—LA PLATA.

| No. of the Observation. | Time of Commencement and Termination of Trial. | Revolutions of the Screw during Trial. | Revolutions per Minute. | Current in Feet per Minute. | Level of Trial during Observation below Surface. |
|---|---|---|---|---|---|
| No. 1 | h.  m.  s.<br>3  30  0 P.M.<br>3  36  2 „ | 522 | 86·5 | 93·68 | 4 feet. |
| „ 2 | 3  38  0 „<br>3  41  5 „ | 207 | 86·5 | 93·77 | 4 „ |
| „ 3 | 3  43  0 „<br>3  46  4 „ | 239 | 77·8 | 84·94 | 7 „ |

TABLE I.—(continued).

| No. of the Observation. | Time of Commencement and Termination of Trial. | Revolutions of the Screw during Trial. | Revolutions per Minute. | Current in Feet per Minute. | Level of Trial during Observation below Surface. |
|---|---|---|---|---|---|
| No. 4 | h. m. s.<br>3 48 0 P.M.<br>3 50 58 „ | 259 | 87·2 | 94·37 | 4 feet. |
| „ 5 | 3 53 30 „<br>3 56 30 „ | 210 | 70·0 | 77·11 | 10 „ |
| „ 6 | 3 58 45 „<br>4 1 45 „ | 266 | 88·7 | 95·88 | 4 „ |
| „ 7 | 4 4 0 „<br>4 7 3 „ | 157 | 51·3 | 58·34 | 16 „ |
| „ 8 | 4 9 30 „<br>4 12 33 „ | 279 | 91·5 | 98·69 | 4 „ |
| „ 9 | 4 20 0 „<br>4 21 0 „ | 38 | 38·0 | 44·98 | 1 foot from bottom. |
| „ 10 | 4 25 0 „<br>4 27 0 „ | 63 | 31·5 | 38·49 | „  „ |
| „ 11 | 4 30 0 „<br>4 32 0 „ | .. | .. | .. | „  „ |
| „ 12 | 4 37 0 „<br>4 38 0 „ | 47 | 47·0 | 54·02 | „  „ |
| „ 13 | 4 42 0 „<br>4 43 0 „ | 38 | 38·0 | 44·98 | „  „ |
| „ 14 | 4 48 0 „<br>4 50 0 „ | 69 | 34·5 | 41·47 | „  „ |
| „ 15 | 4 54 0 „<br>4 55 0 „ | 98 | 98·0 | 105·21 | 4 feet below surface. |
| „ 16 | 4 56 0 „<br>4 57 0 „ | 99 | 99·0 | 105·22 | „  „ |
| „ 17 | 4 58 0 „<br>4 59 0 „ | 87 | 87·0 | 94·17 | — |
| „ 18 | 5 0 0 „<br>5 1 0 „ | 82 | 82·0 | 89·15 | 10 „  „ |
| „ 19 | 5 6 0 „<br>5 11 0 „ | 382 | 76·4 | 83·53 | Surface to bottom and bottom to surface, three times in succession. |
| „ 20 | 5 14 0 „<br>5 19 3 „ | 495 | 97·6 | 104·81 | 4 feet below surface. |

The above Table summarizes the result of the current observation on the La Plata on the 30th December, 1870. We are now prepared to represent the result by diagrams; and these will show what figures alone do not reveal without graphic representation. The result must, however, be considered in conjunction with the tide records, in order that the whole subject may be understood. We have accordingly represented on the diagrams all observations on tide, wind, barometer, and thermometer, which had been registered in the

Book of the Tide Records for the 30th December, 1870, and part of the 31st, and we will now proceed to explain the diagrams, and analyze the result of the observations by the light which they throw on the subject.

## DIAGRAMS.

On Plate II. we have three diagrams, all of which refer exclusively to the La Plata, its currents, tides, winds, &c., as obtained by observation on the 30th December, 1870. Diagram No. 2 is a summary, and represents the situation and the circumstances as a whole; Diagrams No. 1 and No. 3 are details, arising from No. 2, by representing on a much larger scale what happened at certain hours of the day. We shall first consider Diagram No. 2.

In this diagram the tidal rise and fall for every quarter of an hour for the whole day is given, commencing from midnight December 30th, and continuing to noon of next day. The first line to the left of the diagram represents midnight, and every succeeding hour is shown by a line two-tenths of an inch from the preceding one; accordingly, the length of one day is equal to a space of $4\frac{2}{10}$ inches. To the first line corresponding to midnight a scale is attached divided into feet and inches, representing the tide gauge; from the position of the surface of the La Plata on the gauge as observed and registered in the Book of Tide Records, the outline of the wave is determined by marking every quarter of an hour on the diagram the height of the La Plata according to the scale of the diagram; for example, at 7 A.M., December 30th, the La Plata marked 6 feet 5 inches on the gauge; and by drawing a horizontal line from the scale of the diagram at 6 feet 5 inches until it intersects the vertical hour line for 7 A.M., we obtain a point of the tidal wave corresponding to that hour of the day; and so on with every hour or quarter of an hour for the whole day. The force of the wind and the rise and fall of the barometer and thermometer are also noted, all of which explain themselves on the diagram, and we only remark that these should be carefully studied before any other part of the diagram be taken into consideration. The meridian passages of the moon are also shown by placing the centre of a disk representing the moon, to the hour and minute of her transit; the first of which appears at 6.35.30 A.M. for the meridian below the horizon to which the top of wave X corresponds, being the succeeding high water. It is the time of neap tides, the moon being at her first quarter, and she happens moreover to be in her greatest southern declination, which accounts for the great relative difference in the height of the two waves. The upper transit of the moon occurs at 6.56 P.M. The shaded part of the diagram represents night, from 6 P.M. to 6 A.M. The commencement and the termination of the current observation appear on the wave X during flood to ebb tide; and it so happens that these observations were made at half tide, a fortunate circumstance, not that the time could not have been determined beforehand from preceding tide observations, but that in such situations the necessary preparatory arrangements are little under control; and still more fortunate was it that there was no disturbance of any kind from wind, also shown by the regular outline of the wave as determined from the tide records. Looking at the shaded part M resting on the back of wave X, and which represents the whole time of current observations, from 3.30 to 5.20 P.M., and which on account of the small scale of the diagram is compressed to within ¾th inch width, we nevertheless note at a glance that the current represented by a heavy line intersecting the horizontal one marked 100, is increasing from the commencement of

**trials, and reaches** its maximum just before their termination, and that the maximum **current registered** by the meter corresponds to the maximum inclination of the tidal wave, **determined by** independent observations on shore. These, and many other similar coincidences of observations carried on a hundred miles apart, and independent of each other, **have** over and over again proved the accuracy and the unerring truthfulness and fidelity **of** current meter observations conducted with a little care and attention. There will be ample evidence in this respect in the following pages.

The scale representing velocity of currents in **feet per minute is arbitrary, and to keep** the diagrams within reasonable limits we **assumed the inch to represent 30 feet distance** traversed by the current in **one minute; and throughout our observations we took the** minute as the **standard of time for comparison. It is a convenient space of time, and may** frequently be **taken as the time of observation, whilst the second is impracticable,—too** small a fraction **to be accurately measured; it may be taken with equal accuracy as a** standard for **calculation to represent results, but the time of observation must always** extend over many seconds, and should not be much under a minute. The nature of the observation determines the time and the means to be employed, and we now refer particularly to currents.

Diagram No. **3** represents on a large scale the current observations shown on the back of the tidal wave X in the small space of ⅔th inch width, the horizontal scale being magnified twelve times, and which permits every observation being shown in its exact position as to time; the vertical scale remaining in both diagrams Nos. 2 and 3 the same. The enormous exaggeration of inclination of the surface of the La Plata, as shown by the tidal wave X from flood to ebb, is a little reduced in Diagram **No. 3,** but it still **remains very great, as we** shall note subsequently. In **this diagram commencing at 3 P.M. and terminating at 5 P.M., the position** of the **surface of the La Plata is again shown as observed on the gauge during those hours; the vertical** scale of the diagram **is the same as that of No. 2, the horizontal scale being only enlarged; the** space representing **one hour in** Diagram **No. 2** represents **only five minutes in No. 3.** With these premises we proceed to consider Diagram **No. 3 illustrating the** preceding Table, containing a summary of the result of the current observations on the La Plata on the 30th December, 1870.

Observation No. 1 falls between 3.30 and 3.36 P.M.; the current observed is the mean during that space of time supposing it had changed, and it would **belong to 3.33 P.M.,** for which hour and minute it should be plotted. **In the present case it is, however,** immaterial whether the current be **put down for 3.30 or for 3.36, because it was constant as shown by the subsequent trial, and the curve representing the** velocity would remain **the same; but the rule should be adhered to whether or not a** deviation from it would **materially affect the result; it is not** more trouble to do it correctly. Drawing a vertical **line corresponding to 3.33 P.M.** on the diagram, and marking on that line from a datum assumed anywhere—in the present case, assumed to be at the low-water line—a distance equal **to the** velocity of current in feet per minute, viz. 93·68, by a scale representing feet, **we obtain a** point marked No. 1 on the diagram. By a similar proceeding **we obtain** point No. 2 on a line 3h. 39m. 32s. P.M.; No. 3 on **a line at 3h.** 44m. 32s., **with** a corresponding velocity of 84·94 feet, and so on for every succeeding number a corresponding point at the precise time of the trial is obtained for the velocity of current on a certain level, the latter being also noted to the point.

If we now connect the points by a continuous line which represents velocities on the same level, we obtain a curve which in a graphical way represents the current at that depth during those hours. If the line should rise, the current increases; if it should fall, it decreases; if it should be parallel to the datum, it is constant; and by the curve we see not only that the current is increasing or decreasing which the figures also show, but we perceive at a glance the mode of the rise or fall, which the figures do not disclose to the eye. Connecting all the points of velocities 4 feet below surface, we obtain the curve of velocity of current for that depth, on the diagram represented by a hard line, taken as a datum for comparison of other velocities at other levels. This curve shows, that at the commencement the current was constant for about twenty minutes; that it then began slowly to increase, the rate being greatest at about 4.30 P.M., when, still continuing to increase at a smaller rate, the maximum current was reached at about 5.5 P.M., from which time it commenced again to decrease.

On the other levels, we have only at the lowest 1 foot from the bottom of the La Plata a succession of observations, and these seem to oscillate to and fro within certain limits. If we take the mean between the first two on this level, viz. the mean between Nos. 9 and 10, and further take the mean between the three points on the same level terminating these observations, viz. Nos. 12, 13, and 14, and construct these geometrically by taking the centre of gravity of the first two and also of the latter three points, and connect the two centres so found by a straight line, we obtain a line a b, approximately representing the increase of the current at that level. We say approximately, because from these trials it would appear that the current is not as steady near the bottom as it appears in the upper levels; the line will represent the mean of the currents at a given time, oscillating right and left of a b on the diagram. The other points due to other levels are isolated, and do not throw light on the question if considered in the mode adopted for the points on the 4-foot level and those near the bottom of the La Plata, but these isolated points, considered in a different manner, are of great interest and value for the science of hydraulic engineering.

Suppose a section of the La Plata were to be made in the direction of the current at 3.56 P.M., we should have a group of observed points at various levels close to that line of section, some preceding it a little, others following it; and since at that time the current of the La Plata was nearly constant, we should in that section have the position of various points of velocities belonging to various depths; and, although slight alterations may and will have taken place, these changes cannot materially affect the result, and no corrections need be attempted on that account. Following this line of argument, Diagram No. 1 arises. This diagram is drawn on a still larger scale and is not distorted, horizontal and vertical scale being the same, viz. 1 inch to 10 feet. The current is from the Paraná to the Atlantic, on the diagram from left to right; and the position of the meter lowered and raised from the platform in a vertical line, is shown on the diagram by A B.

At 3.56 P.M., the current 4 feet below surface is obtained by Diagram No. 2, from the curve representing velocity of currents, as determined by the observations. This current differs but little from that of the preceding and subsequent trials, and its value is ascertained from that diagram at 95 feet 2 inches per minute, which, marked on a line 4 feet below surface, determines a point as the position of a drop of water on that level carried by the current that distance from line A B in one minute. Proceeding in a similar manner with the observations made on the 7, 10, and 16 foot levels, *viz.* by plotting the currents in feet per minute of Nos. 3, 5, and 7, on lines 7, 10, and 16 feet below surface, all of which commence at the vertical line A B, we obtain a series of terminal points, each representing a drop of water carried by the current on that level in one minute a distance of 84 feet 11 inches, 77 feet 1 inch, and 58 feet 4 inches respectively. If we further prolong the line *a b*, representing approximatively the velocity of the current 1 foot from the bottom of La Plata until it intersects the line of section for 3.56 P.M. on Diagram 2, we obtain the current on that level for that time at 35 feet 0 inch per minute, and which, plotted in a similar manner as the preceding currents, will give another terminal point of a drop of water carried on that level a distance of 35 feet from A B in one minute.

We have now five terminal points on five different levels, each of these indicating the distance from line A B the current has travelled in one minute on each level : and if we connect by a straight line the point on the 4-foot with that on the 16-foot level, being two observations the greatest distance apart, and prolong it in both directions, we find that the other points belonging to other levels are nearly intersected, and that the five points spreading over the whole distance from near the surface to near the bottom, are practically in a straight line, having an inclination of 17° 50' with the horizon. One minute previous these five points were in a vertical line, namely, within A B, its inclination being then 90° 0' with the horizon ; it appears, then, that in open channels of uniform depths, in which the current is not disturbed and free to follow the inclination of the surface, all the particles of water which at one time are within a straight line may be considered to remain in a straight line the inclination of which is changing, and that the currents from surface to bottom decrease at a uniform rate.

The law may not be universal, and further trials must determine the point ; but certain it is that the currents of the La Plata were governed by that law on the 30th December, 1870. The observations on the 30th December, as shown in Table I. as a summary of results, also enable us to try the accuracy and truthfulness of this conclusion by check observations of a different nature and by a different mode of proceeding ; and these substantially confirm the above law of the movement of currents in the La Plata.

The first section was made for 3.56 P.M. and referred to currents in force at that time, or very near it. We may, however, prepare another section belonging to currents at 5.8 P.M. ; and here we have the mean current from surface to bottom with great accuracy, and we know besides the currents 4 and 10 feet below surface by observation, and the current 1 foot from bottom by construction. If we should possess no other observation except that of the mean current obtained by the integrator, we should have a line which must intersect the curve representing the terminal points of all currents then in force, which moved the particles of water from A B a certain distance on each level in one minute ; and, whatever that curve may be, a parabola, an ellipse, or any other sinuous curve, the line obtained for the mean current must so intersect and divide it, that the area of all the points projecting in advance of the mean line must be equal to the area of all the

G

points of the curve which are in the rear of it. The integrator determined the mean current at 5.8 P.M. to be 83·53 feet from surface to within 1 foot of the bottom of La Plata. If we now draw a vertical line 83 feet 6 inches distant from A B, this line must so intersect the curve of the terminal points of the various currents then in force —whatever it may be—that the sum of all the currents which exceed that line must be equal to the sum of all the currents which fall short of it; and if we halve the vertical mean line, we may draw through this mean centre any straight line which will do the same thing as the vertical one, viz. it will intersect the curve of terminal points in such manner that the excess of one side of it will be balanced by the deficiency on the other side; excess and deficiency referred to the new line. All this rests upon the mean current observation alone.

We have, moreover, observed independently the terminal points of currents on three different levels, viz. on the 4, 10, and 23 foot levels, the latter being 1 foot from the bottom of the La Plata. If we now connect the point of the 4 and the 23 foot level by a straight line, we find, that this line passes right through the mean centre, and also near the point of the 10-foot level. Any line passing through the mean centre is, however, a mean line; accordingly these points which form the curve, fall on a straight line which at the same time is a mean line; and, if other points on other levels not observed should not also fall on the same line, they must oscillate to the right and left of it, and excess on one side cover deficiency on the other. The distribution of the independent points is however such, that there is not much room left for oscillations round the mean line, which could only be by a very sinuous curve of double curvature, because the terminal points of the curve are on the mean line, and besides another point about half-way the depth; and, should all the terminal points of the curve not simultaneously fall into the mean and oscillate to the right and left of it, the straight line would nevertheless properly represent them.

We come here to a conclusion by a process and an argument very different from that employed in the construction of the first section for 3.56 P.M.; and yet the conclusion in both cases is substantially the same; and we may now safely say that the currents of the La Plata decrease from surface to bottom in a direct ratio governed by the distance from surface to bottom.

The difference in the velocity of currents of the sections corresponding to 3.56 P.M. and 5.8 P.M. is but small; the one section being near the minimum, the other near the maximum during the period of observations; the difference amounts to 10 per cent. on the 4-foot level, and yet, with this small difference in the increase of the current, the diagrams begin to reveal and to indicate another law, viz. that the increase in the velocity of currents grows more rapidly at the bottom than at the surface; for, the inclination of the line of terminal points of currents is greater with the section for the greater current at 5.8 P.M. In the first section it was 17° 50', in the second it is 18° 47'; this is also shown by the figures; the current 4 feet below surface increased from 95 feet 2 inches to 104 feet 8½ inches, or 10 per cent. in round figures; the current about 23 feet below surface (1 foot from the bottom) increased from 35 feet to 50 feet per minute, or 43 per cent., and this happened at the same point, with a depth a little reduced; the increase of velocity of current being due to an increase of inclination of the surface of the La Plata. The observations are not numerous enough, nor is the change in the current great enough, to follow this question further at present; we only call attention to points which these observations determine and to those which they indicate, as matters for further investigation.

## MEAN CURRENT.

The mean current may be ascertained with great accuracy by the current integrator, and its correct measurement is usually of greater importance to the engineer than that of any other current at or below the surface. Whenever the inquiry of an engineering question involves volumes of water, the mean current alone determines the point; and it is only the exception that an engineering problem is not directly or indirectly affected by the volumes which a certain section may pass in a given time. We consider mean current observations of primary importance. The next current of importance is that near the bottom of a channel. It is frequently a vital question in reference to the execution of works, the maintenance of channels, &c., and should always be observed next to the mean current. The surface current is of interest, but usually of not much importance. On account of convenience it is frequently observed, and taken as a standard for comparison. The convenience, however, is more apparent than real; for, no current can be observed with accuracy by the use of surface floats; and with a current meter an observation near the surface or many feet below may usually be made with equal convenience, especially in rivers of moderate size. Wherever systematic observations are made, three currents should be observed, viz. mean, bottom, and surface; the third as a check on the preceding two; because any two of the three currents determine the third, as shown in the preceding pages.

The observations of the 30th December include only one mean current trial near half tide, when the current of the La Plata was near its maximum. We made, however, a number of observations on the proceding day, viz. the 29th December in the afternoon, on a tidal wave corresponding to that of the 30th, both being due to the lower transit of the moon. These observations include three consecutive mean current trials, again terminating them by a trial on the 4-foot datum level; and we will give the details of these observations by a copy of the entries in the Survey Book.

" DECEMBER 29TH, 1870, AFTERNOON. LOCALITY—OUTER ROADS."

"Calm weather, gentle current of air from the north. Surface slightly undulating. Meter No. 1 attached to Integrator. Observations from platform resting on two boats 12 ft. apart."

| No. of the Observation. | Time of Commencement and Termination of Trial. | Time of Observation. | Indices of Meter. | Position of Meter. | Remarks. |
|---|---|---|---|---|---|
| No. 20 | h. m. s. 5 11 0 P.M. 5 15 20 „ | m. s. 4 20 | 14a + 125 16a + 48 | Surface to bottom. | Meter twice lowered and raised in succession. |
| „ 21 | 5 19 0 „ 5 25 15 „ | 6 15 | 16a + 48 18a + 100 | Surface to bottom. | Meter three times lowered and raised in succession. |
| „ 22 | 5 27 0 „ 5 35 33 „ | 8 33 | 18a + 100 21a + 105 | Surface to bottom. | Meter four times lowered and raised in succession. |
| „ 23 | 5 40 0 „ 5 45 5 „ | 5 5 | 21a + 195 27a + 188 | 4 ft. below surface. | Meter stationary. |

"Depth of water under platform, 22 ft. at 5.20 P.M."

"Direction of current to the Atlantic. Magnetic bearing of current by prismatic compass, 135° looking in the direction of current."

G 2

If we now proceed to work out the preceding entries copied from the Survey Book, in the manner adopted for the observations of the 30th December, we obtain a summary of results which will appear from the following Table :—

TABLE II.

DECEMBER 29TH, 1870.  OUTER ROADS, LA PLATA.

| No. of the Observation. | Time of Commencement and Termination of Trial. | Revolutions of the Screw during Trial. | Revolutions per Minute. | Current in Feet per Minute. | Level of Trial during Observation below Surface. |
|---|---|---|---|---|---|
| No. 20 | h. m. s. 5 11 0 5 15 20 | 315 | 72·70 | 79·8₂ | Surface to bottom for Mean current. |
| „ 21 | 5 19 0 5 25 15 | 454 | 72·64 | 79·7⁶ | „      „      „      „ |
| „ 22 | 5 27 0 5 35 33 | 606 | 71·11 | 78·2₂ | „      „      „      „ |
| „ 23 | 5 40 0 5 45 5 | 485 | 95·41 | 102·6₁ | 4 feet below surface, Stationary. |

The mean current trials Nos. 20, 21, and 22 are remarkable ; and if any doubt should linger with anyone as to the accuracy and truthfulness of the records of the current integrator, these observations should dispel that doubt.   We have here three consecutive observations within 30 minutes, in a channel within which from flood to ebb the current changes but little within hours, and in a short space of time it changes next to nothing ; and the trials, each extending over a considerably different period of time, and performed over various distances by the meter, give, when worked out, practically identical results.

In trial No. 20, the meter was lowered and raised from surface to bottom twice in succession, travelling in each down and up journey 21 feet, together 42, in both 84 feet ; within this distance it was exposed to all the various currents then in force during 4 minutes and 20 seconds, reckoning the time from the moment of immersion to the instant at which it was raised out of the water, viz. the instant it appeared the second time above the surface ; the result of the observation being a mean current of 79 feet 9₁₀⁶ inches per minute.

In the trial No. 21, immediately succeeding the former, the same operation of lowering and raising is performed three times in succession, the meter travelling in all over a distance of 126 feet, exposed to the various currents then in force during 6 minutes and 15 seconds ; and the result of the observation is a mean current of 79 feet 9 inches per minute, differing from the former ₁₀⁶ of an inch ; the current slightly decreasing.

Then follows trial No. 22, in which the same operation is performed, excepting that the distance over which the meter was made to travel is further increased, namely, to 168 feet, the time consumed being 8 minutes and 33 seconds.  The result of the observation is a mean current of 78 feet 3 inches per minute, further decreasing a little, viz. 1½ foot per minute since the last observation.

A single trial may be most accurate or it may be most inaccurate, there is nothing to show; a check trial, however, decides the question. But it also depends how the check trial is made. Suppose the meter did not record the true current, that it omitted or only partially recorded the smaller currents, and that by letting the meter pass twice the distance occupied by the weak currents, it had in each passage lost 10 feet from the true mean current, and that the result of No. 20 trial should therefore be 20 feet short of the true mean. If we now made a check observation repeating in every respect the first trial, and if the meter performed its mechanical operation in both cases in an identical manner, we should have the same result in the same current, and so far the check observation would be satisfactory; yet it would leave us entirely in the dark whether the current shown by the meter is the true mean or not; for, the check trial would be also 20 feet short for the same reason as the main trial had been. But if we extend the distance, and by the check observation travel three times over the distance of the weak currents, the mean would have been short three times 10, or 30 feet; and since the first trial only fell 20 feet short, and the check trial by the latter mode of observation 30 feet, we should have a difference of 10 feet in the result of mean currents between the two, independent of the meter performing its mechanical operation in an identical manner. If, however, the result remains the same, no matter whether the meter travels once, twice, or three times over the distance of weak and strong currents, it will not only be proof that the meter performed its mechanical operation in a satisfactory manner, doing duty equally, but that it also gave full and proper value to the strong and to the weak currents, and that the result obtained by the meter for the mean current is the true mean under observation.

The above trials of mean currents established our confidence in the current meter, and determined us to abide by its records as long as we have proof that it is in perfect adjustment, however unexpected or inexplicable the records of the meter may at times have appeared.

In the following Table, III., we give a summary of the current observations especially in reference to the mean current, and we also add surface and bottom currents.

### TABLE III.

#### CURRENTS OF THE LA PLATA.

| Date of Observation, December, 1870. | No. of Trial. | Mean Current in Feet per Minute. | Surface Current in Feet per Minute. | Bottom Current in Feet per Minute. | Current 4 ft. below Surface in Feet per Minute. | Percentage of Mean Current referred to Surface Current. |
|---|---|---|---|---|---|---|
| 29th, 5.15 P.M. | No. 20 | 79·82 Obsn. | 118·6 Consn. | 41·1 Consn. | 103·39 Consn. | 67·90 per cent. |
| „ 5.25 „ | „ 21 | 79·76 „ | 118·5 „ | 41·0 „ | 103·36 „ | 67·30 „ |
| „ 5.35 „ | „ 23 | 78·22 „ | 118·0 „ | 38·5 „ | 101·61 Obsn. | 66·29 „ |
| 30th, 5.11 „ | „ 19 | 83·53 „ | 116·75 „ | 50·0 Curve. | 104·81 „ | 71·54 „ |
| „ 3.56 „ | „ (a) | 71·50 Section. | 108·0 „ | 35·0 „ | 95·40 Curve. | 66·20 „ |

*Abbreviations.*—Obsn., for Observation; Consn., for Construction.

In Table III. we have a summary of observations which we propose to analyze. With each current it is stated how it was obtained, whether by direct observation or from the curve resulting from observations, or by construction, following the geometrical proceeding which the trials themselves determined, and which has been fully discussed, page 40 and 41.

The figures of Table III. are perplexing ; and we were by no means prepared for such results. The trials of the 29th December, Nos. 20, 21, and 22 show nothing unusual, and bear one another out; we have almost identical currents throughout; a gentle decrease of the mean, followed by nearly similar decreases of the surface, bottom, and 4-feet-below-surface currents. All this is in accordance with generally accepted principles. So the current observations of the 30th December, if considered by themselves, present nothing unusual; with the latter there is a much wider range in the magnitude of the various currents, but they follow the generally accepted rules; the decrease from No. 19 to No. a has a different ratio in the various currents, strongly marked, and with some currents somewhat unexpected.

If we, however, compare the results of the 29th December with those of the 30th (the observations were made in the same locality on both days, under identical circumstances of weather), we seem to fall into contradictions and such deviations from the generally accepted rules on currents, that we begin to doubt whether these results can possibly be correct; and, if we had not cumulative proof of their individual and intrinsic accuracy—not only an approximation within the units, but accuracy to within the second decimal—we should not have attempted to trace the cause of these conflicting and apparently irreconcilable results.

The decrease in the superficial current on the 29th December is trifling, the units are the same; the decrease is only visible in the first decimal, and all will admit that the superficial current is remarkably uniform. If we compare No. 22 of the 29th December with No. 19 of 30th, we find by commencing at the surface, that the current was 118 feet per minute with the former and 116⅔ with the latter, that it decreased 1 foot 3 inches per minute; 4 feet below surface it had, however, increased 2 feet 2 inches; and near the bottom it had increased 11 feet 6 inches; in short, a decrease in the surface current was attended by a considerable increase in the bulk of all the currents below surface, which is further demonstrated by the increase of 5 feet 4 inches per minute of the mean current; and this is a large increase in the opposite direction when referred to the change at the surface. The result is counter established or rather accepted rules. It is against the rule that we should have in the same locality, under similar circumstances, with a decreasing surface current, a great increase in the mean current. The result would still remain unaccountable, even if we drop the difference in the surface currents, and assume that both were alike; we should still have for similar surface currents in the same locality without any visible external disturbance a great difference in the mean current, the increase being 7 per cent. on the preceding mean.

There is further conflicting evidence; even if we assume the surface current identical we should have the mean current expressed as a percentage of that on the surface to vary from 61¼ to 71½ per cent.; whilst on the same days other trials also demonstrate that the percentage may remain the same, viz. 66⅓ per cent. for two surface currents differing considerably, differing most among the observations; for, we have in trials No. 22 and No. a the same percentage for a mean current belonging to a surface current of 118 feet and to another 108 feet per minute.

These results at once explode the generally accepted rule that the mean current of a river may be found by taking so many per cent., say 60, 70, 80, or more of the superficial current. The rule is erroneous; the mean current cannot be so expressed. These trials seem to strike at the root of our knowledge on the subject.

If we consider the currents near the bottom in reference to those at the surface, we find by comparing No. 22 and No. 19 that with a slight decrease at the surface, we have a great increase at the bottom; and if we again drop the difference in the surface currents and take them to be alike, we have still the unaccountable fact, that for some reason or other independent of the surface velocity, the current near the bottom increased 30 per cent.; and we shall be able to show that it may change a great deal more with the same surface current. Where are the rules, which for a certain current at the surface give a definite one at the bottom? the Tables, which for every surface current give the corresponding velocity at the bottom of a channel? and do not currents most materially affect the question of maintenance of channels?

The ready way and the convenient way to get over the difficulty is to say that the observations cannot be correct, and that some error has crept in, producing the contradictory results at variance with established rules. This is the way to dispose with a high hand of matters which declare our ignorance. Others, more considerate, may say, that the differences are too small to determine the point either way; at any rate, not large enough to call into question accepted principles of hydraulic science. With the former there is no argument. We remind the latter that the cradle of the discovery of Great Truth and the greatest laws of Nature is the Differential Calculus; the consideration of changes as a whole of interdependent quantities brought about by infinitesimal changes of some one of its quantities. To comprehend and to appreciate the above conflicting evidence, let us apply the unerring principles of the differential calculus as far as they may be applied to observations, where infinitesimal changes must be coarsely represented by small differences still measurable with accuracy by instruments and other means employed to obtain observations.

The current originates by gravitation, the mass of water following the inclination of its surface. Inclination is therefore one of the main quantities involved in the problem. Movement having set in, the current increases until resistance prevents further acceleration, and the current becomes at the locality a uniform quantity. Resistance is therefore the next quantity involved, and it originates essentially from the bottom of the channel, and appears as friction between a fluid and a solid. As yet we know nothing about the inclination of the La Plata; it is clear that it can only be small, and we know that the superficial velocity strongly marks the slightest alteration in the inclination or in the fall of the surface, especially when the total is but a small quantity. There is a slight decrease in the superficial current from No. 22 to No. 19, and still smaller must be the decrease in the inclination of the surface during the latter trial; and, since the difference in the surface currents does not amount to more than about 1 per cent., the inclination may have differed but very little, if at all, and could never account for an increase of 30 per cent. in the current near the bottom.

As to the next quantity, viz. the resistance of the bottom, both trials having been made at the same locality, not more than one day apart, the resistance could not have changed; and having had calm weather, with a gentle breeze from the north right across the line of current, the resistance of the air was both insignificant and similar in each case, and could in no way account for the changes in the currents of No. 22 and No. 19. These conclusions are perplexing enough, and though "differentiation" cleared the way, it only tells us to seek the cause elsewhere.

Having found that there can be no change either in the inclination of the surface or in the resistance of the bottom, and considering that the mean current was increased from 78·22 to 83·53 feet per minute, there must necessarily have been greater power during the time of the higher mean current, overcoming that resistance at a higher speed, and especially if we consider that the resistance itself was augmented considerably by the very circumstance of the higher velocity of the mean current. Where is this greater power to come from? Inclination alone is no power, but combined with a mass free to follow gravitation it may constitute power. Inclination and the force of gravitation having remained the same, the greater power could only have arisen from a greater mass. We come therefore to the conclusion that during observation No. 19, apart from increased velocity, greater masses must have been in motion than during No. 22; and since the trials were made on the same spot, the greater mass affected by the inclination of the surface could only have arisen from an increased depth of water.

If we now refer to the entries of the Survey Book, we do find increased depth, although the entries cannot determine the point, the figures having reference to different hours of the day. The tidal observations on shore, however, confirm the above argument by the records of levels corresponding to the day and hour of the observation, and determine the difference of level to the inch; according to those records the surface of the La Plata on the 29th December at 5.35 P.M. was 20 inches lower than on the 30th December at 5.11 P.M., from which it follows, that during the trial (No. 19) with the greater mean current, there was also 20 inches greater depth of water.

Depth alone is, then, enough by itself to change all the currents, and materially to change them. The difference of depth between No. 22 and No. 19 is only about ⅟₄₄th part, and yet there is a little revolution in the currents; the great majority are increased, some remain the same, others near the surface are decreased. And it is not difficult to see why depth alone should bring about all these changes; and since we consider this to be a vital question, we shall more fully analyze it in order that the fact may not only be accepted but also be clearly understood.

A mass of water follows the inclination of its surface by the force of gravitation. If it met no resistance it would continue to accelerate and its velocity would be every moment greater. As it meets resistance which increases with velocity, a point will be reached at which the two opposing forces balance, and acceleration will cease and the velocity of the mass will at the locality remain a constant quantity. The resistance to the movement of water following its declivity within a channel arises from the bottom alone, and appears as friction between a solid and a liquid. It is the bottom of a channel which opposes the movement of every particle of water within it, and it is plain that the particles nearest to the bottom are subject to the greatest resistance; those farther from it to less resistance, because the farther the particles be the more indirectly are they acted upon by the retardation of adjoining particles. It would follow that whatever the velocity of the current be at the bottom, it must in some manner increase with the distance until the surface be reached where the maximum velocity would be attained.

It should be, however, clearly understood, that the whole resistance originates only with the bottom by retarding the fluid in contact with it, and that it is due to the reaction of the particles of water on each other that a variety of currents from bottom to surface are produced; that the primary cause of the resistance is the retardation of the particles of water on the bottom; and that the "measure" of that resistance is the friction between the liquid and the solid, expressed by the velocity of the current in the immediate proximity of the bed of the channel. All the other currents are an effect produced by the reaction of that collision and friction, and they are no "measure" of the resistance encountered. We are here opening a dark page in the book of hydraulic science, and we must ask for patience and special attention. Our distinguished authors argued the point ingeniously, and traced the resistance to the difference of currents in two adjoining levels, and to the consequent friction between the particles of the liquid. They had but few reliable facts to guide them, and in their theory many took the effect for the cause. The resistance is not caused by the friction of the liquid particles in adjoining levels; but the friction of the liquid particles in adjoining levels, arising from a difference of velocity, is the effect of the resistance offered by the bottom to the movement of the liquid. The distinction is material; it is essential.

That the resistance originates entirely by friction between solid and liquid, we have the proof, that the moment we release the liquid from the solid, by assuming the friction to be nil, all the differences in the currents of the various levels disappear, and we have but one current from surface to bottom, the whole mass accelerating and moving freely through space.

By moving the mass of the liquid from a higher to a lower level, the work of gravitation is balanced by the work of resistance between the solid bottom and the liquid mass; and that resistance is the equivalent of it, no matter what the intermediate currents may be from surface to bottom. The work of Gravitation depends on two quantities—viz. distance and mass, both having equal voice in the result, and the product of the two represents the work of gravitation. As long as that product remains the same (either quantity may vary to any extent), the work of the Resistance between the liquid and the solid will also be similar—in short, the current near the bottom will remain the same.

The distance may be represented by the fall per mile; the mass, at the locality under consideration, may be expressed by the depth and the mean current within that depth. We have accordingly the following fundamental general equation :—

(Current at bottom) = function { (fall per mile) (mass of water) } ;

or, at a certain locality, within a vertical line :—

(Current at bottom) = function { (fall per mile) (mean current) (depth) } .

H

From the equation it appears that the bottom current will at a given locality depend on three different quantities, and that its expression is rather involved. We might at once proceed to express this law by a formula and determine the "function," but this is not our object for the present; above all, the subject-matter under consideration should be clearly understood; cause and effect and the mode of change should be fully comprehended; the rest is a mere formality.

The preceding arguments also led us to the conclusion that, whatever the current at the bottom may be, its velocity must increase with the distance, and that the maximum current should be at the surface. This conclusion is fully confirmed by trials which will be matter for subsequent investigation. We already know that on the La Plata the increase from bottom to surface is in direct proportion to the distance, expressed by a straight line; but whether it be a straight line or a curve of some sort, it does not affect our present argument; all that now concerns us is the fact, that the reaction on the liquid particles arising from the resistance of the channel decreases with the distance; that its effect is less and less felt by the currents farther and farther from the bottom. Let us assume we had the same current at the surface in two localities where the depth is different; what must happen? The effect of the resistance of the channel had been in both cases transmitted to the surface where it appeared similar. Since, however, the same effect was felt at the greater distance, the cause in operation—the resistance of the bottom—must have been greater where the depth was greater; because if it had been the same, the effect at the increased distance would have been smaller, manifesting itself by a smaller surface current. It is therefore certain that, whatever may be the law of increase or decrease, with similar surface currents the corresponding currents at the bottom will be different with different depth, and that in the greater depth there must be the greater bottom current. This fundamental law is independent of the mode of increase of currents from bottom to surface; the current at the surface will, however, depend on the law of increase.

These considerations show the fallacy and the error of certain accepted rules of hydraulics; such as making the bottom current depend on the surface current, as if these quantities were only depending on each other; as if for a certain surface current of say 300 feet per minute, a bottom current of 228 feet per minute would correspond; the bottom current may be anything within the limits of nil and 300 feet per minute. Equally erroneous is the rule which determines the mean current at so many per cent. of the surface current; the mean current may at the same locality be anything within the limits of 50 to 100 per cent. of the surface current; and the error is committed by making the one only to depend on the other, as if no third or fourth quantity had a voice in the result. The mean current of the La Plata is the arithmetical mean between the currents at the surface and at the bottom. For the same mean current a variety of surface and bottom currents may correspond, the only condition to be fulfilled is, that half the sum of the two should remain the same quantity. The mean current of a river at a given locality is a primary quantity, of the same importance as its fall and depth; the surface and bottom currents are secondary quantities, of which the latter is the more important.

## INCLINATION.

As an estuary of the sea, the inclination of the La Plata should on the average be nil. The tidal wave would constantly change the inclination of its surface at a given locality from zero to a positive and in turn to a negative maximum, making the currents flow correspondingly in opposite directions. The great rivers however interfere, and complicate the question of the fall of the La Plata, which, by the change of the inclination, becomes positive and negative.

Let us first consider these changes on the La Plata, assuming for a moment that the great rivers discharged their volume elsewhere. Under these circumstances the changes in the inclination of the surface may be readily determined by tidal observations at two stations a certain distance apart. If the channel of the estuary be regular, all the features of the tidal waves will reproduce themselves in outline at the two stations, although at different times ; and the difference in the time will depend on the distance between the two stations. The gauge which records the outline of the wave, only measures the level of the surface at the time ; and we find that a certain level at one station makes its appearance at the other station so many hours and minutes later ; and that if we consider the position of the surface at the same instant of time, the one is so many feet higher than the other, which for the time being will be the fall between the two stations ; and with the distance and the difference of level we have the fall per mile at that locality at a given time.

The tidal wave is said to travel at so many miles an hour ; and it means this : that the position of the entire wave shifted so many miles an hour, retaining its outline. If, for example, the two stations were 20 miles apart, and all the succeeding features of the wave made their appearance at the other station exactly one hour later, we should say that the tidal wave travelled in the estuary at a rate of 20 miles per hour, inasmuch as the entire wave shifted its position 20 miles in that space of time. The rate at which a tidal wave is propagated in an estuary being known from observation, the tidal records at one station will determine the inclination of the surface of the estuary at that locality, because we know the rise or fall of the surface in one hour, and we also know the distance at which the surface due to that rise or fall stood one hour previous ; and, accordingly, we know the inclination at a given time. We can, however, by a gauge only determine the inclination of the surface on waves of translation ; and the wave having passed, and the consequent rise and fall of the surface having ceased, the gauge will record a permanent position of the surface, which may be a true level, or it may be an inclination, the gauge does not further disclose. This may readily be seen by observing a gauge within a river of great inclination. If the flow be constant and nothing should interfere with the discharge of the river, its level will remain constant, and the gauge will record a permanent level, although the inclination of the river's surface may be very great ; here, it does not disclose more than if it were placed on a pond. The gauge determines the inclination of the surface during the passage of the tidal wave, if nothing should interfere with its formation, and the surface of the water be only affected by the wave. As soon, however, as the rise or fall of the surface may be produced by other causes jointly with the translation of the wave, the gauge no more discloses the inclination of the surface.

E 2

The La Plata is an estuary of the sea, in which we have the translation of a tidal wave affecting its surface, and the discharge of great rivers also affecting it. For a time the tidal wave and the rivers act in the same way; the one does not interfere with the other. During another time, however, the effects of the tidal wave and that of the rivers are opposed to each other. If there were no tide, the volume discharged by the rivers into the La Plata would determine the inclination of its level, which for a constant volume would remain the same, and the rivers by themselves would cause neither a rise nor a fall in the surface of the estuary. The approach of the tidal wave, however, reverses the current in some parts, in others it reduces the current to zero, the wave acting as a bar on the rivers; which, in consequence, by themselves begin to raise the surface of the La Plata, independently of the tide. It follows that under these circumstances the rise, as shown by the gauge, is not caused by the tidal wave alone, but is jointly the effect of the tide and the rivers; and we can no more determine the inclination of the surface from the tide gauge, because we cannot tell how much of the rise belongs to the tide and how much to the rivers. Under these circumstances the rise of the surface of the La Plata during the approach of the tidal wave must be more rapid, greater in the same space of time.

If, for example, the rise due to tide should at a certain locality be 5 inches per hour, which means that a point of the wave 5 inches higher than the level at the gauge would in one hour reach the gauge and raise the surface 5 inches; and, if the supply from the rivers should, by a reduction of the current at the point one hour distant from the gauge, be large enough to fill the space between the former level of the gauge and that point 5 inches higher by the difference between supply and discharge, we should have the surface rising on account of the rivers as fast as on account of the tide; and between the gauge and the point of the wave one hour distant a true level would be maintained, although the rise of 5 inches divided by the distance, would give a fall from the approaching wave to the rivers. Should the rivers, by the tide checking the current a certain distance below, raise the surface of the La Plata faster than the tide, a declivity will be still maintained from the rivers to the approaching wave, and not from the wave to the rivers, as indicated by the gauge, which would be the reverse of the fact. Should the wave raise the surface faster than the rivers, a declivity will be from the wave to the rivers; but in all these cases the gauge cannot by any means be made to determine the inclination, which no more depends on its records.

It is different, however, with the second half of the wave, when its declivity and its currents correspond with those of the rivers in direction. Here the rivers will neither increase nor decrease that inclination; their effect being only to prolong the duration of the wave by the supply of volumes in addition to those which belong to the wave proper. From flood to ebb the records of the gauge may therefore be used to determine the inclination of the surface of the La Plata. The wave X on Diagram No. 2, Plate II., is particularly steady in its descent, the weather being most favourable, and the regularity of the descent itself shows that there was nothing to disturb the uniform flow from flood to ebb. On the back of this wave the current observations were made about half tide; and on the sister wave, preceding it 24 hours, current observations were made which extended over the greater part of the wave, and particularly at its termination the observations were extended over a considerable period at low water, the level of the La Plata being stationary. During the current observations of the 30th December, from 3.20 P.M. to 5.20 P.M., the tidal wave propagated itself in this reach of the river, as determined by another series of observations, at a rate of 12 miles per hour. The fall in the surface of the La Plata from 3.45 P.M. to 5.15 P.M., on the same day, was from 5' 2" on gauge to 4' 7", or equal to 7 inches, which divided by the distance corresponding to 1¼ hour, will give a fall of 0·388 inch per mile as the mean fall of the La Plata during the current observations. In a similar manner, the maximum fall during the current observations on the 30th December was determined at 0·444 inch per mile; and the minimum fall from high to low water at 0·342 per mile, all on the same day for the wave X from flood to ebb.

These figures for the inclination of the surface of the La Plata from flood to ebb are so small, and the variations so trifling, that we should have hesitated to accept them had we not proof of their accuracy by a very different mode of proceeding, in which the rate of propagation of the tidal wave does in no manner enter into the question, and does not form part of the solution of the problem. They belong to a tidal wave of great regularity from flood to ebb, and during a fall of its surface for nine hours; and the inclination was the same nearly for the whole estuary. During the next five hours from ebb to flood, the inclination will not only constantly have varied in the same locality, but the degree of variation will have been different in every locality; and all that may be said is, that in the lower reaches of the estuary the inclination will have been reversed to − 0·342 inch per mile; at others it will have been for a time 0·000; and at localities near the mouths of the rivers it will have been a little under + 0·342 inch per mile. The mean inclination of the La Plata during 24 hours is therefore less than 0·342 inch per mile, and since the bulk of the discharge of the volume of the great river takes place in the greater depths, it follows that if no tide existed and the surface of the La Plata were stationary, its inclination would be a little greater than the mean level.

It is an interesting question, what would be the inclination of the surface of the La Plata when the tidal wave had passed away, and before the next wave had commenced raising its surface again, viz. during a permanent level at low water. As a rule, the level does not remain constant for more than a quarter to half an hour, but occasionally it remains constant for more than an hour. We happen to possess observations which, commencing on the wave, extended a considerable distance over low water, then lasting an hour. We were during the current observations not aware that we had at the time a constant level corresponding to low water; these trials determine the inclination of the La Plata when not affected by tides. The observations were made on the 29th December, 1870, in the afternoon, and have already been referred to as the mean current trials in this chapter.

On Diagram No. 2, Plate II., we have shown the outline of wave V on which these trials were made. This wave belongs to the preceding day, and joins wave W at ⊙. The position of V on the diagram is that belonging to it on the 29th December. The details of the observations are given on page 43, and we shall only consider their result. Close at the end of wave V the surface current was nearly the same as on the subsequent day at half tide for the maximum current at 5 P.M., for which we found the fall to be 0·444 inch per mile; and the current remained within a trifle constant during low water, decreasing only about half foot per minute. The depth of water was, however, on the 29th a little less, so that the inclination must have been a little greater, which will fall between $\frac{1}{2}$ and $\frac{7}{12}$ inch per mile. These figures represent the inclination of the La Plata in that reach of the estuary when free from tidal influence.

It thus appears that the moment the tidal wave has passed the gauge, the latter records a constant level, although the inclination of the surface of the La Plata remains the same; just what would happen in an ordinary river of constant level, the surface of which may have any inclination. It also appears, that the inclination of the tidal wave from flood to ebb and that of the La Plata when undisturbed by tides, are very nearly parallel; and that from flood to ebb the surface of the La Plata falls by degrees parallel to itself, until it reaches low water, when it remains permanent, parallel to its former positions.

## CHAPTER IV.

## The Paraná de las Palmas.

### THE RIVER.

THE Paraná de las Palmas is one of the two main branches of the Paraná in the delta of the main river. About 60 miles from the La Plata the main stream divides in two branches—the Paraná de las Palmas and the Paraná Guazú—which for a short distance convey together nearly the whole volume of the Paraná; the latter branch, however, soon divides into a number of smaller ones, forming a great number of islands between the Palmas branch and the Uruguay. The Palmas does not further divide, and conveys by one channel its whole volume into the La Plata, receiving on its course a few insignificant tributaries.

On the accompanying chart, Plate I., the great rivers and the La Plata are shown on two different scales. Chart No. 1, on the smallest scale, shows the general configuration of the great rivers Paraná and Uruguay, and of the La Plata estuary. Chart No. 2 represents on a large scale the upper end of the La Plata estuary, in front of Buenos Ayres, and the locality of the current observations in the Outer Roads, and also the lower portion of the delta of the Paraná, with the locality of some sections, including that of the Palmas, which will form the subject of this chapter.

On the 16th January, 1871, we sailed at 9 A.M. from the Tigre through the Arroyo del Capitan into the mouth of the Paraná de las Palmas; and, ascending the river, we steamed at a slow pace, watching the course of the Palmas and the formation of its banks. The channel of the Palmas is not irregular, and the bends are as a rule gentle; rapid turns are comparatively few. Throughout its whole length both its banks are low-lying marshy land a couple of feet above the level of the river; at mean water level perhaps 3 feet higher than the river, and the whole land right and left appears to be thickly covered with a coarse grass of long growth, from 6 to 12 feet high, and occasionally with forests of a peculiar tree called "Seivo"—which seems to monopolize these wild districts of hundreds of square miles of area, with here and there some willows and poplars; and, although the scenery is pleasing and will make an agreeable impression on all who first view it, its uniformity soon makes it monotonous.

The Palmas is deep throughout, and of easy navigation. The difficulties of the navigator only commence at its mouth as he enters the La Plata; not only because the depth is much reduced, but also that the channel of the Palmas is no more visible, being submerged in the waters of the La Plata; and although a few buoys would remove the principal difficulty, such luxuries are as yet little known in South America. To avoid the embouchure of the

Palmas, steamers and other vessels take a tortuous narrow channel called the Arroyo del Capitan, which leads them to the embouchure of a small river called "Lujan," which, though obstructed by a short bar inconveniently shallow, may be navigated without buoys. Passengers from Buenos Ayres are usually taken by rail to San Fernando at the mouth of the Lujan; and, embarking within the bar, the vessels sail by the "Capitan" into the Palmas, and take their up-journey on these magnificent rivers. On the right bank of the Palmas, the main land of the province of Buenos Ayres is occasionally approached, which may be from 40 to 50 feet higher than the level of the river, and is different in appearance from the land forming its banks. Wherever the main land approaches the Palmas, a settlement, a village, or even a little town, occupies the shore as near as practicable to the river, which is the highway of communication.

Late in the evening on the 16th the 'Aguila' dropped anchor a little above the junction of the Baradero with the Palmas, the former being a small branch of the main river. The Baradero is a favourite with the small craft, on account of its gentle current and the shorter distance, avoiding many inconvenient turns at the head of the Palmas, and because of the shelter which the main land affords along its course. The weather was calm and very fine for observations, although exceedingly hot. On the morning of the 17th the locality for a section was selected; its position is approximately shown on the chart, and marked "Palmas Section." We had previously explored the Palmas, Guazú, and other branches of the main river, with a view to ascertain a number of localities favourable for observations; and to select the best and most convenient for the purpose of the survey. Having determined the line of section, measured the base, fixed the main triangulation points, and having moored an observatory about midway the river and placed a gauge at its margin, we were in the evening prepared to commence systematic observations early next morning.


THE TIDE.

We proceeded to the head of the Palmas to get, if possible, out of tidal reach. We could not obtain any information as to the fall of the great rivers—nothing which might even indicate it; and supposing the fall of the Palmas to be something like that of other great rivers, like that of the Mississippi for example, amounting to a couple of inches per mile determined by extended surveys of that river—and, considering that the ordinary rise of the La Plata was under 4 feet from low to high water, we should have expected that from 60 to 70 miles up the Palmas the tide of the La Plata would not have been perceptible; at any rate, not affecting materially the currents.

As soon as the line of section was determined upon, a gauge was put down at the margin of the river, and observations on that gauge commenced at noon on the 17th January. From noon to 5 P.M. the gauge fell steadily 5½ inches; it then commenced rising again steadily; and we find the following entry in the Survey Book :—

"From 5 P.M. the river commenced slowly rising again; it seemed to be affected by the tide of the La Plata; to be ascertained by observations, whether the rise be tidal, and whether it affects the velocity of current.

"Temperature at noon in the shade, 108° Fahr.; at 6 P.M., 93° Fahr."

This was an inconvenient discovery; for, if the tide should affect the current, it would be difficult, if not impracticable, to determine the volume of the Palmas with an approach to accuracy in a moderate space of time, on account of variation in the tidal rise; nor would the object be gained if the trials extended over many days, if simultaneous observations could not be made on the other great branch, the Paraná Guazú; and we had but one steamer fitted out for the purpose; and nothing could be done in this wilderness without a complete equipment of men, provisions, and instruments.

It was, however, still uncertain whether the couple of inches of rise and fall of the surface of the water at the Palmas section was due to tide, or to some other disturbance occasioned by wind or by oscillations of the surface of these great rivers, which in many places look more like seas than rivers. Having, moreover, all the preparatory arrangements completed for systematic observations, we determined to try the question; and, in view of a possible tidal interference, to extend the observations day and night without interruption of any kind, commencing the trials at five o'clock in the morning of the 18th January, 1871. We shall now proceed to give the entries in the Survey Book which refer to the observations and measurements at the Palmas section.

## DIARY OF OBSERVATIONS.

"LOCALITY: PARANÁ DE LAS PALMAS, JANUARY 17TH, 1871, 2 P.M."

Base line on left bank, 300 feet long by steel tape, magnetic bearing of base looking up river, 338° 20'; the base on the margin of river; fixed flag on right shore at right angles with the base.

Line of base 13 feet 5 inches from margin of river, bank vertical about 3 feet above level of water.

Observatory moored about half-way the river, on line of section.

Angles: From Observatory on base, A B, 28° 42'.

   ,,    From flag O on right shore on A B, 13° 38½'.

Current at Observatory: 5.15 P.M., January 17th, 1871, preliminary observations:

Current Meter No. 1: 5 min. 0 sec., Observation, Meter $(5a + 23) - 0 = 1028$ revolutions; or, per minute 205½ revolutions.

Check observation: 5.18 P.M., 3 min. 0 sec., Observation, Meter $(8a + 40) - 1028 = 620$ revolutions; or, per minute, 206½ revolutions.

These preliminary observations were made when the river was lowest during the day, measuring 1' 6½" on gauge, the river commenced slowly rising again.

JANUARY 18TH, 1871. LOCALITY: PARANÁ DE LAS PALMAS, AT OBSERVATORY MOORED ON RIVER. CURRENT METER No. 1, 6 A.M. CALM WEATHER.

*Surface Current at Observatory.*

| No. | | | Current observation | | | | | | | Meter | |
|---|---|---|---|---|---|---|---|---|---|---|---|
| No. | 1. | 6.0 A.M. | Current observation | 5 min. 0 sec. | | | | | | Meter | $(2a + 162) = 564$ revolutions. |
| | | 6.5 ,, | ,, | ,, | 1 | ,, | 0 | ,, | | ,, | $\begin{cases}(2a + 162) \text{ commencement.}\\(3a + 73) \text{ termination.}\end{cases}$ |
| ,, | 2. | 7.30 ,, | ,, | ,, | 1 | ,, | 0 | ,, | | ,, | $(0a + 138) = 138$ revolutions. |
| | | 7.33 ,, | ,, | ,, | 2 | ,, | 0 | ,, | | ,, | $\begin{cases}(0a + 138) \text{ commencement.}\\(2a + 16) \text{ termination.}\end{cases}$ |
| ,, | 3. | 1.33 ,, | ,, | ,, | 1 | ,, | 0 | ,, | | ,, | $(0a + 170) = 170$ revolutions. |
| | | 8.36 ,, | ,, | ,, | 2 | ,, | 0 | ,, | | ,, | $\begin{cases}(0a + 170) \text{ commencement.}\\(2a + 100) \text{ termination.}\end{cases}$ |
| ,, | 4. | 9.32 ,, | ,, | ,, | 1 | ,, | 0 | ,, | | ,, | $(0a + 176) = 176$ revolutions. |
| | | 9.35 ,, | ,, | ,, | 2 | ,, | 0 | ,, | | ,, | $\begin{cases}(0a + 176) \text{ commencement.}\\(2a + 136) \text{ termination.}\end{cases}$ |
| ,, | 5. | 10.30 ,, | ,, | ,, | 1 | ,, | 0 | ,, | | ,, | $(0a + 185) = 185$ revolutions. |
| | | 10.35 ,, | ,, | ,, | 5 | ,, | 0 | ,, | | ,, | $\begin{cases}(0a + 185) \text{ commencement.}\\(5a + 114) \text{ termination.}\end{cases}$ |

I

| No. 6. | 11.33 A.M. | Current observation | 1 min. 0 sec. | Meter | $(0a + 192) = 192$ revolutions. |
| | 11.35 „ | „ „ | 5 „ 0 „ | „ | $\{(0a + 192)$ commencement. |
| | | | | | $\{(5a + 182)$ termination. |
| „ 7. | 12.30 „ | „ „ | 1 „ 0 „ | „ | $(0a + 196) = 196$ revolutions. |
| | 12.33 „ | „ „ | 5 „ 0 „ | „ | $\{(0a + 196)$ commencement. |
| | | | | | $\{(5a + 151)$ termination. |
| „ 8. | | Interim observation simultaneously with No. 1 and No. 2 Meters | | | |
| | 12.20 P.M. | No. 1 Meter | 5 min. 0 sec. observation | | $(4a + 140) = 944$ revolutions. |
| | 12.20 „ | „ 2 „ | 5 „ 0 „ „ | | $(5a + 73) = 1078$ „ |
| | 12.50 „ | „ 1 „ | 5 „ 0 „ „ | | $(4a + 180) = 984$ „ |
| | 12.50 „ | „ 2 „ | 5 „ 0 „ „ | | $(5a + 107) = 1112$ „ |
| „ 9. | 1.30 „ | Current observation | 1 „ 0 „ | Meter | $(0a + 199) = 199$ revolutions. |
| | 1.35 „ | „ „ | 5 „ 0 „ | „ | $\{(0a + 199)$ commencement. |
| | | | | | $\{(5a + 189)$ termination. |
| „ 10. | 2.30 „ | „ „ | 1 „ 0 „ | „ | $(1a + 0) = 201$ revolutions. |
| | 2.35 „ | „ „ | 5 „ 0 „ | „ | $\{(1a + 0)$ commencement. |
| | | | | | $\{(5a + 180)$ termination. |
| „ 11. | | Interim observation. Comparative of superficial velocity and that 3 feet below surface. | | | |
| | 3.20 „ | Surface: Current observation 2 min. 0 sec. Meter No. 1 | | | $(1a + 186) = 387$ revolutions. |
| | 3.25 „ | 3 ft. below: „ „ 2 „ 0 „ „ „ | | | $(1a + 178) = 379$ „ |
| | | Surface velocity 8 revolutions higher in two minutes than 3 feet below surface. | | | |
| „ 12. | 3.30 „ | Current observation | 1 min. 0 sec. | Meter | $(0a + 191) = 191$ revolutions. |
| | 3.35 „ | „ „ | 5 „ 0 „ | „ | $\{(0a + 191)$ commencement. |
| | | | | | $\{(5a + 144)$ termination. |
| „ 13. | 4.0 „ | Interim observation. Memorandum: At 4 P.M. the surface current was 197 revolutions, the result of several observations. | | | |
| „ 14. | 4.30 „ | Current observation | 1 min. 0 sec. | „ | $(0a + 200) = 200$ revolutions. |
| | 4.32 „ | „ „ | 5 „ 0 „ | „ | $(5a + 2)^* = 1007$ revolutions. |
| † „ 15. | 5.47 „ | „ „ | 1 „ 0 „ | „ | $(0a + 200) = 200$ revolutions. |
| | 5.49 „ | „ „ | 5 „ 0 „ | „ | $\{(0a + 200)$ commencement. |
| | | | | | $\{(5a + 196)$ termination. |
| † „ 16. | 6.30 „ | „ „ | 1 „ 0 „ | „ | $(0a + 200) = 200$ revolutions. |
| | 6.32 „ | „ „ | 5 „ 0 „ | „ | $\{(0a + 200)$ commencement. |
| | | | | | $\{(5a + 180)$ termination.** |
| † „ 17. | 7.30 „ | „ „ | 1 „ 0 „ | „ | $(0a + 192) = 192$ revolutions. |
| | 7.32 „ | „ „ | 5 „ 0 „ | „ | $\{(0a + 192)$ commencement. |
| | | | | | $\{(5a + 152)$ termination. |
| † „ 18. | 8.30 „ | „ „ | 1 „ 0 „ | „ | $(0a + 193) = 193$ revolutions. |
| | 8.32 „ | „ „ | 5 „ 0 „ | „ | $\{(0a + 193)$ commencement. |
| | | | | | $\{(5a + 150)$ termination. |
| | | Memorandum: The notations of observations No. 15, 16, 17, and 18, marked with a cross, fell into the river, but were immediately recorded by the assistants in charge from memory. | | | |
| „ 19. | 9.30 „ | Current observation | 1 min. 0 sec. | „ | $(0a + 188) = 188$ revolutions. |
| | 9.32 „ | „ „ | 5 „ 0 „ | „ | $\{(0a + 188)$ commencement. |
| | | | | | $\{(5a + 95)$ termination. |
| „ 20. | 10.30 „ | „ „ | 1 „ 0 „ | „ | $(0a + 175) = 175$ revolutions. |
| | 10.32 „ | „ „ | 5 „ 0 „ | „ | $\{(0a + 175)$ commencement. |
| | | | | | $\{(5a + 54)$ termination. |
| „ 21. | 11.30 „ | „ „ | 1 „ 0 „ | „ | $(0a + 165) = 165$ revolutions. |
| | 11.32 „ | „ „ | 5 „ 0 „ | „ | $\{(0a + 165)$ commencement. |
| | | | | | $\{(5a + 85)$ termination. |

Steamer 'Capitan' passes, Observation No. 21 doubtful.

* The meter set to zero at commencement of trial.     ** Figure 180 not certain.

### January 19, 1871.

| No. 22. | 0.30 A.M. | Current observation | 1 min. 0 sec. | Meter | $(0s + 187) = 187$ revolutions. |
| | 0.32 ,, | ,, ,, | 5 ,, 0 ,, | ,, | $\begin{cases}(0s + 187)\text{ commencement.}\\(5s + 74)\text{ termination.}\end{cases}$ |
| ,, 23. | 1.30 ,, | ,, ,, | 1 ,, 0 ,, | ,, | $(0s + 180) = 180$ revolutions. |
| | 1.32 ,, | ,, ,, | 5 ,, 0 ,, | ,, | $\begin{cases}(0s + 180)\text{ commencement.}\\(5s + 113)\text{ termination.}\end{cases}$ |
| ,, 24. | 2.30 ,, | ,, ,, | 1 ,, 0 ,, | ,, | $(0s + 187) = 187$ revolutions. |
| | 2.32 ,, | ,, ,, | 5 ,, 0 ,, | ,, | $\begin{cases}(0s + 187)\text{ commencement.}\\(5s + 159)\text{ termination.}\end{cases}$ |
| ,, 25. | 3.0 ,, | ,, ,, | 1 ,, 0 ,, | ,, | $(0s + 196) = 196$ revolutions. |
| | 3.5 ,, | ,, ,, | 5 ,, 0 ,, | ,, | $\begin{cases}(0s + 196)\text{ commencement.}\\(5s + 150)\text{ termination.}\end{cases}$ |

Memorandum : At 3 A.M. weather a perfect calm, the river like a mirror, reflecting every star.

| ,, 26. | 3.30 A.M. | Current observation | 1 min. 0 sec. | ,, | $(0s + 194) = 194$ revolutions. |
| | 3.35 ,, | ,, ,, | 5 ,, 0 ,, | ,, | $\begin{cases}(0s + 194)\text{ commencement.}\\(5s + 157)\text{ termination.}\end{cases}$ |
| ,, 27. | 4.30 ,, | ,, ,, | 1 ,, 0 ,, | ,, | $(0s + 194) = 194$ revolutions. |
| | 4.35 ,, | ,, ,, | 5 ,, 0 ,, | ,, | $\begin{cases}(0s + 194)\text{ commencement.}\\(5s + 142)\text{ termination.}\end{cases}$ |
| ,, 28. | 5.30 ,, | ,, ,, | 1 ,, 0 ,, | ,, | $(0s + 170) = 170$ revolutions. |
| | 5.35 ,, | ,, ,, | 5 ,, 0 ,, | ,, | $\begin{cases}(0s + 170)\text{ commencement.}\\(4s + 192)\text{ termination.}\end{cases}$ |
| ,, 29. | 6.0 ,, | ,, ,, | 1 ,, 0 ,, | ,, | $(0s + 155) = 155$ revolutions. |
| | 6.5 ,, | ,, ,, | 5 ,, 0 ,, | ,, | $\begin{cases}(0s + 155)\text{ commencement.}\\(4s + 180)\text{ termination.}\end{cases}$ |
| ,, 30. | 6.30 ,, | ,, ,, | 1 ,, 0 ,, | ,, | $(0s + 136) = 136$ revolutions. |
| | 6.35 ,, | ,, ,, | 5 ,, 0 ,, | ,, | $\begin{cases}(0s + 136)\text{ commencement.}\\(4s + 1)\text{ termination.}\end{cases}$ |
| ,, 31. | 7.40 ,, | ,, ,, | 1 ,, 0 ,, | ,, | $(0s + 100) = 100$ revolutions. |
| | 7.43 ,, | ,, ,, | 5 ,, 0 ,, | ,, | $\begin{cases}(0s + 100)\text{ commencement.}\\(2s + 191)\text{ termination.}\end{cases}$ |
| ,, 32. | 8.30 ,, | ,, ,, | 1 ,, 0 ,, | ,, | $(0s + 97) = 97$ revolutions. |
| | 8.33 ,, | ,, ,, | 5 ,, 0 ,, | ,, | $\begin{cases}(0s + 97)\text{ commencement.}\\(2s + 187)\text{ termination.}\end{cases}$ |
| ,, 33. | 9.0 ,, | ,, ,, | 1 ,, 0 ,, | ,, | $(0s + 99) = 99$ revolutions. |
| | 9.5 ,, | ,, ,, | 5 ,, 0 ,, | ,, | $\begin{cases}(0s + 99)\text{ commencement.}\\(2s + 187)\text{ termination.}\end{cases}$ |
| ,, 34. | 10.0 ,, | ,, ,, | 1 ,, 0 ,, | ,, | $(0s + 105) = 105$ revolutions. |
| | 10.4 ,, | ,, ,, | 5 ,, 0 ,, | ,, | $\begin{cases}(0s + 105)\text{ commencement.}\\(3s + 31)\text{ termination.}\end{cases}$ |
| ,, 35. | 11.0 ,, | ,, ,, | 1 ,, 0 ,, | ,, | $(0s + 142) = 142$ revolutions. |
| | 11.5 ,, | ,, ,, | 5 ,, 0 ,, | ,, | $\begin{cases}(0s + 142)\text{ commencement.}\\(4s + 38)\text{ termination.}\end{cases}$ |
| ,, 36. | 12.0 noon | ,, ,, | 1 ,, 0 ,, | ,, | $(0s + 161) = 161$ revolutions. |
| | 12.4 P.M. | ,, ,, | 5 ,, 0 ,, | ,, | $\begin{cases}(0s + 161)\text{ commencement.}\\(4s + 145)\text{ termination.}\end{cases}$ |
| ,, 37. | 12.30 ,, | Interim observation | | ,, | 172 revolutions per minute. |
| ,, 38. | 1.10 ,, | Current observation | 1 ,, 0 ,, | ,, | $(0s + 180) = 180$ revolutions. |
| | 1.15 ,, | ,, ,, | 5 ,, 0 ,, | ,, | $\begin{cases}(0s + 180)\text{ commencement.}\\(5s + 63)\text{ termination.}\end{cases}$ |
| ,, 39. | 2.0 ,, | ,, ,, | 1 ,, 0 ,, | ,, | $(0s + 172) = 172$ revolutions. |
| | 2.4 ,, | ,, ,, | 5 ,, 0 ,, | ,, | $\begin{cases}(0s + 172)\text{ commencement.}\\(5s + 48)\text{ termination.}\end{cases}$ |
| ,, 40. | 3.6 ,, | ,, ,, | 1 ,, 0 ,, | ,, | $(0s + 185) = 185$ revolutions. |
| | 3.8 ,, | ,, ,, | 5 ,, 0 ,, | ,, | $\begin{cases}(0s + 185)\text{ commencement.}\\(5s + 101)\text{ termination.}\end{cases}$ |
| ,, 41. | 4.2 ,, | ,, ,, | 1 ,, 0 ,, | ,, | $(0s + 186) = 186$ revolutions. |
| | 4.4 ,, | ,, ,, | 5 ,, 0 ,, | ,, | $\begin{cases}(0s + 186)\text{ commencement.}\\(5s + 107)\text{ termination.}\end{cases}$ |

I 2

No. 42.                                    Interim observation.  Trial for Mean Velocity of current.

4 26  0 P.M.  Commencement }  5 min. 38 sec.  {Meter No. 1, three times down and three times up in
4 31 38   „   Termination   }                 { succession.

Indexes of Meter {Commencement (0a +  0)} = 856 revolutions,
                 {Termination   (4a + 52)}

on which followed immediately an observation for surface velocity, centre of meter
12 inches below surface; observation by wire:

4.33  „    Current observation  1 min. 0 sec.  Meter (0a + 177) = 177 revolutions.

Followed by an observation for velocity of current at bottom, centre of meter 1 foot
from bottom; observation by wire:

4.36  „    Current observation  1 min. 0 sec.  Meter (0a + 122) = 122 revolutions.

„ 43.                                     Interim observation.  Trial for mean velocity of current.

4 46  0  „   Commencement }  4 min. 50 sec.  Meter No. 1, three times in succession down and up.
4 50 50  „   Termination   }

Indexes of Meter {Commencement (0a +  0)} = 629 revolutions.
                 {Termination   (3a + 26)}

Followed immediately, observation for surface velocity by wire:

4.52  „    Current observation  1 min. 0 sec.  Meter (0a + 156) = 156 revolutions.

„ 44.   5.9  „    Current observation  1 min. 0 sec.  Meter  (0a + 161) = 161 revolutions.
        5.11  „      „        „       5  „  0  „      „   {(0a + 161) commencement.
                                                         {(4a + 153) termination.

---

JANUARY 19TH, 1871.  LOCALITY: PARANÁ DE LAS PALMAS.

*Currents on Line of Section across River.*

No. 45.  12.30 P.M.   At Observatory.  Meter No. 1 continued.
                      Angle on A B, 28° 42'  Meter, 1 min. 0 sec. observation  170} = 172 revolutions.
                                                                               174}

„ 46.  12.35  „    From small boat:
                   Angle on A B, 35° 33'       „  1  „  0  „      „     148 revolutions.
                   After first observation:
                   Angle on A B, 35° 45'       „  1  „  0  „      „     153   „

„ 47.  12.51  „    From small boat:
                   Angle on A B, 47° 58'       „  1  „  0  „      „     144   „
                   After first observation:
                   Angle on A B, 47° 58'       „  1  „  0  „      „     150   „

„ 48.  1.2   „     From small boat :
                   Angle on A B, 61° 47'       „  1  „  0  „      „     134   „
                   After first observation:
                   Angle on A B, 61° 30'       „  1  „  0  „      „     136   „

„ 49.  1.7   „     At Observatory :
                   Angle on A B, 28° 42'       „  1  „  0  „      „   (0a + 182) = 182 revolutions.
                                               5  „  0  „      „   {(0a + 182) commencement.
                                                                  {(5a +  63) termination.

„ 50.  1.30  „     From small boat:
                   Angle on A B, 21° 30'       „  1  „  0  „      „     131 revolutions.
                   After first observation:
                   Angle on A B, 21° 30'       „  1  „  0  „      „     135   „

„ 51.  1.40  „     From small boat :
                   Angle on A B, 16° 47'       „  1  „  0  „      „      84   „
                   After first observation :
                   Angle on A B, 16° 38'       „  1  „  0  „      „      84   „

MEMORANDA ON CURRENTS. *January 19th, 1871.*

"Since 5 P.M. the river suddenly commenced rising very fast, viz. from 5 P.M. to 6 P.M. it rose 12 inches; it continued rising fast, though less rapidly, until at 9 A.M. it stood 3 feet 8 inches on gauge, a rise of 26 inches in four hours.

"The river was like a mirror at 9 P.M., reflecting the stars, much like a pond. Observations having been discontinued since 5 P.M., intending to start to another locality for a section, we thought it desirable to see what the current was, expecting it to be very feeble. Left 'Aguila' for the permanent observatory midway the river with meter No. 1, to observe current. Meter immersed 1 minute 0 second indicated nil; next observation meter immersed 2 minutes 0 second. Indexes not changed, indicating nil. The night being a perfect calm, and the current, if any, being too small for the meter, with paper wetted and floating on the surface as a sheet, we found a slight current towards Obligado of about 4 feet in 30 seconds. The Paraná de las Palmas reversed ! flowing towards its sources."

*January 19th, 1871.*—LOCALITY: PARANÁ DE LAS PALMAS.

*Soundings on Line of Section.*

From board the Steamer 'Aguila.'

January 19th, 6.45 P.M., soundings commenced.

| | | | |
|---|---|---|---|
| No. 1. | Angle 47° 28' | Depth 40 feet 5 inches. | Gauge 2 feet 10½ inches. |
| „ 2. | „ 71° 10' | „ 20 „ 7 „ | |
| „ 3. | „ 32° 35' | „ 48 „ 8 „ | |
| „ 4. | „ 17° 35' | „ 42 „ 7 „ | |
| „ 5. | „ 21° 55' | „ 54 „ 2 „ | Gauge 3 feet 3 inches. |

Sounding No. 5 at 7.40 P.M.

January 20th, 7.15 A.M.

| | | |
|---|---|---|
| No. 6. | Angle 28° 42' at Observatory between boats on platform : | |
| | First sounding, 52 feet 5 inches | At 7.15 A.M. |
| | Second sounding, 52 „ 6 „ | |
| | Gauge on shore, 4 „ 0 „ | |

| | | | |
|---|---|---|---|
| No. 7. | Angle 62° 10' | Depth 36 feet 8 inches. | At 8.15 A.M. |
| „ 8. | „ 41° 43' | „ 43 „ 5 „ | |
| „ 9. | „ 27° 40' | „ 53 „ 8 „ | |
| „ 10. | „ 20° 26' | „ 53 „ 0 „ | |
| „ 11. | „ 17° 54' | „ 45 „ 3 „ | |
| „ 12. | „ 14° 44' | „ 5 „ 0 „ | At 9.4 A.M. |

Memoranda on Soundings, January 20th, 8 A.M. :

From C to A, depth 3' 6'' 55 feet from C at 8 A.M.

Depth at C, 3 feet 0 inch at 8 A.M., January 20th.

Eighty feet from C. Depth rapidly increasing; at 120 feet no bottom with 10-feet rod.

MEMORANDUM.

"The above observations complete the section of the Paraná de las Palmas. The river highest when we left, yet a strong current at the time to the La Plata. Velocity of current in the Palmas variable, which decided us to go out of reach of all tides, and to proceed at once as far as Rosario. Left Palmas section for Rosario at 9.4 A.M., January 20th, 1871. Reached the 'Nuevo Vueltas' at 10.13 A.M. Section of Palmas 1h. 2m. from junction of Palmas and Guará branches. Speed of 'Aguila' against shore about 4 miles per hour."

## RISE AND FALL OF RIVER AT SECTION.

Hourly observations were made of the level of the surface of the Paraná de las Palmas at the section by means of a gauge, the position of the surface being noted every hour, day and night. From these records diagrams were prepared similar to those obtained by the tide gauge of the La Plata, and these diagrams determined the position of the surface of the

Palmas at the section for every hour of the day during those observations. Diagram No. 1, Plate IV., shows the rise and fall of the surface of the Palmas by a curve marked "Surface of the Paraná de las Palmas during observations." The same curve is also shown on Diagram No. 1, Plate III., for a longer period. Observations commenced at noon on the 17th January, 1870, and the gauge for the position of the surface of the Palmas at that hour was so placed that it read 2 feet 0 inch, to admit readings a couple of feet below and above that position of the surface. In the diagrams a line was drawn through the lowest level of the Palmas during the observation, occurring on the 19th January, at three o'clock in the morning, which stood at $9\frac{1}{2}$ inches on the gauge in the river. In the diagrams, however, this lowest level of the river was considered the zero of the gauge; to make the entries in the Survey Book referring to the gauge in the river correspond with the gauge of the diagrams, the readings of the river gauge must be diminished by $9\frac{1}{2}$ inches, which will, after the deduction, be the figures for the new datum of the diagram gauge.

## ANALYSIS OF OBSERVATIONS.

We give in the following Tables a summary of the preceding observations on the Paraná de las Palmas, having from the notations as entered in the Survey Books first determined the quantities from the records of the instruments. We have, whenever practicable, two observations on currents immediately following each other, as a check for the accuracy of the trial and that it had been correctly entered in the Survey Book. Referring to the preceding extract, or rather copy, from the Survey Book, we shall by an example or two explain how the quantities had been determined from the notations before we enter into their analysis. Let us take any trial, for example, No. 6 of the surface current obtained at the floating observatory about midway the river. The first trial under that number was a "minute" observation, commencing at 11.33 A.M., the meter being immersed and exposed to the current 1 minute 0 second. The indexes of the meter before trial were set to zero; after the trial the one index (called $a$) had not reached a full division ; the other index stood at 192, the notation being ($o a + 192$) equal to 192 revolutions of the screw of the meter. The minute observation prepares the observer for the sort of current he is about to measure. As soon as practicable, after booking the result of the first, another trial followed at 11.35 A.M., both appearing under No. 6. The second observation extended over a period of 5 minutes 0 second; and, the position of the indexes not having been changed, their reading at the commencement of the second trial was the same as that belonging to the preceding one, viz. ($o a + 192$). After the completion of the second trial the indexes of the meter stood on one wheel ($a$) at 5, and on the other at 132, or (5 $a + 132$), and the value of one division on wheel $a$ being 201, the meter registered accordingly (5 $\times$ 201 + 132) = 1137 revolutions; from which deducting the first reading of 192 belonging to the preceding trial, we have 945 revolutions during the second trial extending over five minutes; and dividing the 945 by 5, we have by the second trial 189 ; which differs from the first by three revolutions.

The same care being taken with each trial, it is obvious that the five-minute trial is nearer the truth than the minute trial ; because there is always an uncertainty as to the exact time the meter was operating by about half a second, namely, the margin of accuracy which an observation may approach. To get an observation of this sort limited to half a second as the maximum deviation, it is necessary to start and terminate the trial within a quarter of a second of the exact time, which is perhaps as much as a good observer may under the circumstances ensure ; if both the deviations from the exact instant of time be of the same sign, *viz.* both positive or negative, and each of the same magnitude, their difference would be nil, and the time elapsed between the two observations would be absolutely correct; if on the other hand they should be of different sign, their difference would be the sum of the two, which would affect the result with the sum of the two deviations ; and if the maximum of each should be a quarter of a second, their sum might be half a second in time ; so that the " minute " observation might be $59\frac{1}{2}$ or $60\frac{1}{2}$ seconds, instead of 60 seconds duration. If the circumstances attending the trial should be less favourable, each deviation might be half a second, which may sometimes neutralize each other ; at others affect the result by their sum, namely, by one second, making the time of the trial 59 or 61 seconds instead of 60.

It follows, that with the same current two consecutive observations may slightly differ, although the observations may be equally accurate, and that the differences will not only differ with the observer, but also according to more or less favourable circumstances. A good observer may under ordinary circumstances determine the time of commencement and termination of trial within a quarter of a second ; the maximum deviation of each trial may then be half a second, plus or minus, and the maximum possible deviation between the succeeding trials may amount to one second. Bearing this in mind, it follows that in a current which will make the screw of the meter revolve three times every second, two consecutive trials, each of one minute's duration, may differ by three revolutions, although the current remained the same, and the observations were equally good. With special trials, and by special care, when the meter is put in and out of gear by pulling and letting go a wire, the observation may be accurate within a tenth of a second, plus and minus, and the result within a fifth of a second certain ; but this will be the possible limit.

By extending, however, the time of observation, we reduce the effect of the error in time ; which may be a quarter of a second more or less than the intended time of the trial. If, for example, the time of trial be five minutes instead of one minute, we may in both cases commence and terminate the observation within a quarter of a second ; but the effect of the error of a quarter of a second on the result of the five-minute observation is only one-fifth of that on the minute trial ; accordingly, with a constant current the five-minute observation will be nearer the truth than the minute trial ; and, on this account, the accuracy of the result will be even greater than usually required. The five-minute observation is therefore the main trial to determine the current per minute, and the minute trial a check, that no grave error was committed ; otherwise there would be no evidence whether the time of the main trial had been four or six instead of five minutes, and a mistake could not be discovered.

As long therefore as the differences between two succeeding trials do not exceed the variation attending good observations, they will confirm each other ; when, however, the differences considerably exceed that limit, it is proof that one or both of the observations cannot be relied on, and should be rejected as "doubtful." Among the preceding trials there is one, No. 21, where the difference between the minute and five-minute observations throws a doubt on the whole trial ; the error may either be in the notation or in the reading of the indexes, or in the time of exposition.

Having found that the "one-minute" observation confirms the "five-minute" trial, both are entitled to consideration, and we may then consider the two trials as one observation extending over six minutes, and by dividing the number of revolutions of the two trials by six we obtain the result with still higher accuracy. This is the course we have followed in the Palmas. In example No. 6, previously quoted, we found the minute observation to give 192 revolutions, the five-minute observation 945, which together make 1137 revolutions for six minutes; and this figure divided by six will give 189½ revolutions per minute as the final result.

The number of revolutions registered by the meter having been found per minute, the value of the revolutions may be expressed in feet by the equation of the meter, see page 36, which will then be the velocity of the current in feet per minute corresponding to that observation.

TABLE IV.

JANUARY, 1871. CURRENTS OF THE PARANÁ DE LAS PALMAS, AT OBSERVATORY MIDWAY THE RIVER.

| Number of Trial. | Mean Time of Observation. | Current in Feet per Minute. | Remarks. |
|---|---|---|---|
| | *January 18th.* | | |
| | h. m. s. | | |
| No. 1 | 6 3 30 A.M. | 119·9 | Centre of meter, 12 inches below surface. |
| „ 2 | 7 32 30 „ | 146·7 | „ „ „ |
| „ 3 | 8 35 30 „ | 174·8 | „ „ „ |
| „ 4 | 9 36 0 „ | 186·8 | „ „ „ |
| „ 5 | 10 35 0 „ | 194·0 | „ „ „ |
| „ 6 | 11 36 30 „ | 197·1 | „ „ „ |
| „ 7 | 12 34 0 P.M. | 209·2 | „ „ „ |
| „ 8 | 12 22 30 „ | 196·4 | „ „ „ |
| | 12 52 30 „ | 203·4 | „ „ „ |
| „ 9 | 1 35 0 „ | 206·6 | „ „ „ |
| „ 10 | 2 35 0 „ | 205·1 | „ „ „ |
| „ 11 | 3 21 0 „ | 201·3 | Centre of meter, 6 inches below surface. |
| | 3 26 0 „ | 197·1 | „ 3 feet 6 inches below surface. |
| „ 12 | 3 35 0 „ | 199·1 | „ 12 inches below surface. |
| „ 13 | 4 0 0 „ | 204·6 | „ „ „ |
| „ 14 | 4 33 30 „ | 208·8 | „ „ „ |
| „ 15 | 5 50 30 „ | 207·8 | „ „ „ |
| „ 16 | 6 33 30 „ | 205·1 | „ „ „ |
| „ 17 | 7 33 30 „ | 200·4 | „ „ „ |
| „ 18 | 8 33 30 „ | 200·1 | „ „ „ |
| „ 19 | 9 33 30 „ | 190·9 | „ „ „ |
| „ 20 | 10 33 30 „ | 184·0 | „ „ „ |
| „ 21 | 11 33 30 „ | ·· | „ „ „ |
| | *January 19th.* | | |
| | h. m. s. | | |
| „ 22 | 12 33 30 A.M. | 187·3 | Centre of meter, 12 inches below surface. |
| „ 23 | 1 33 30 „ | 193·9 | „ „ „ |
| „ 24 | 2 33 30 „ | 201·6 | „ „ „ |
| „ 25 | 3 5 0 „ | 200·1 | „ „ „ |
| „ 26 | 3 35 0 „ | 201·2 | „ „ „ |
| „ 27 | 4 35 0 „ | 198·7 | „ „ „ |
| „ 28 | 5 35 0 „ | 173·5 | „ „ „ |
| „ 29 | 6 5 0 „ | 163·1 | „ „ „ |
| „ 30 | 6 35 0 „ | 141·5 | „ „ „ |
| „ 31 | 7 44 0 „ | 106·0 | „ „ „ |
| „ 32 | 8 34 0 „ | 105·4 | „ „ „ |
| „ 33 | 9 5 0 „ | 105·4 | „ „ „ |
| „ 34 | 10 4 30 „ | 112·9 | „ „ „ |

TABLE IV.—(*continued*).

| Number of Trial. | Mean Time of Observation. | Current in Feet per Minute. | Remarks. |
|---|---|---|---|
| | January 19th. | | |
| | h. m. s. | | |
| No. 35 | 11 5 0 A.M. | 147·7 | Centre of meter, 12 inches below surface. |
| „ 36 | 12 4 30 P.M. | 155·6 | „ „ „ |
| „ 37 | 12 30 0 „ | 179·5 | „ „ „ |
| „ 38 | 1 15 0 „ | 185·5 | „ „ „ |
| „ 39 | 2 4 30 „ | 181·0 | „ „ „ |
| „ 40 | 3 9 30 „ | 191·9 | „ „ „ |
| „ 41 | 4 5 30 „ | 192·9 | „ „ „ |
| | 4 28 40 „ | 153·4 | Mean current, surface to bottom. |
| „ 42 | 4 33 30 „ | 184·5 | Centre of meter, 12 inches below surface. |
| | 4 36 30 „ | 129·3 | Centre of meter, 12 inches from bottom, 49 feet 0 inch below surface. |
| „ 43 | 4 48 25 „ | 137·5 | Mean current, surface to bottom. |
| | 4 52 30 „ | 181·4 | Centre of meter, 12 inches below surface. |
| „ 44 | 5 12 30 „ | 167·0 | „ „ „ |
| „ 44a | 9 0 0 „ | —8·0 | „ „ „ |

TABLE V.

JANUARY 19TH, 1871. PARANÁ DE LAS PALMAS.

*Currents across River on Line of Section.*

| Number of Trial. | Mean Time of Observation. | Current in Feet per Minute corresponding to 1.15 P.M. | Distance of Locality of Observation from A of line of Base. |
|---|---|---|---|
| | January 19th. | | |
| | h. m. s. | | ft. in. |
| No. 45 | 12 30 0 P.M. | 185·5 | 548 0 |
| „ 46 | 12 35 0 „ | 161·4 | 332 3 |
| „ 47 | 12 51 0 „ | 154·4 | 270 5 |
| „ 48 | 1 2 0 „ | 143·4 | 161 11 |
| „ 49 | 1 7 0 „ | 185·5 | 548 0 |
| „ 50 | 1 30 0 „ | 140·3 | 761 7 |
| „ 51 | 1 40 0 „ | 91·2 | 999 5 |

TABLE VI.

*Soundings on Line of Section* (from board the steamer 'Aguila,' January 19th, 1871).

| Number of Sounding. | Depth in Feet and Inches. | Distance of Sounding from A of line of Base. | Time. | Gauge of River. |
|---|---|---|---|---|
| | ft. in. | ft. in. | January 19th. | ft. in. |
| No. 1 | 40 5 | 275 3 | 6.46 P.M. | 2 10½ |
| „ 2 | 20 7 | 102 4 | — | — |
| „ 3 | 48 8 | 469 5 | — | — |
| „ 4 | 42 7 | 946 8 | — | — |
| „ 5 | 54 2 | 745 8 | 7.40 P.M. | 3 3 |
| | | January 20th, 1871. | | |
| „ 6 | 52 5½ | 548 0 | 7.15 A.M. | 4 0 |
| „ 7 | 36 8 | 158 9 | 8.15 „ | — |
| „ 8 | 43 5 | 336 6 | — | — |
| „ 9 | 53 8 | 572 3 | — | — |
| „ 10 | 53 0 | 805 3 | — | — |
| „ 11 | 45 3 | 928 10 | — | — |
| „ 12 | 5 0 | 1140 10 | 9.4 A.M. | 4 2 |

Memorandum : Distance of margin of river from A of base on left bank equal **13 feet 5 inches** ;
A C = 1236 feet 1 inch.

K

Tables IV., V., and VI., giving the result of the various observations on the Palmas in a condensed form, will enable us to prepare diagrams representing in a graphical manner the events registered by the observations; the diagrams will facilitate their comprehension and their analysis; and we believe it would be difficult to arrive at any conclusion on many points without the diagrams, which by their outline and form disclose to the eye that which figures alone would have left unexplained.

## ANALYSIS OF TIDES.

On Diagram No. 1, Plate III., we have shown the rise and fall of the surface of the La Plata due to the tidal wave from the South Atlantic; it will accordingly represent the tide on the La Plata. At the left margin of the diagram we have the tide gauge at Buenos Ayres, and the records on the diagram commence at 10 A.M., January 17, 1871, the surface of the La Plata at that time standing at 4 feet 2 inches on the gauge. The position of the surface is noted every quarter of the hour, and marked on the diagram. So for example, the level remained stationary till 10.15 A.M.; at 10.30 A.M. it fell one inch, and remained at that level till 10.45; at 11.0 A.M. it fell another inch, standing 4 feet 0 inch on the gauge; and so on for every succeeding quarter hour of the day and night the position of the surface of the La Plata is determined by the gauge records. The space on the diagram representing an hour is $\frac{1}{15}$ inch; it is arbitrary, and so taken that the diagram may be kept within a convenient size. We shall, however, see that the outline of the tidal wave itself, its natural form, is by the choice of the vertical and horizontal scales, enormously exaggerated; the tidal waves appearing as a succession of abrupt upheavals of the surface, and there is not the slightest resemblance left between the true form of the wave and as it appears on the diagram, though its outline is correct for engineering purposes and argument.

The first tidal wave of the La Plata marked on the diagram, made its appearance at the gauge of Buenos Ayres at 11.15 A.M., January 17, the level of the La Plata standing at 4 feet 0 inch, being low water. The wave raised the surface of the La Plata to 5 feet 11 inches at 2.30 P.M.; it remained stationary for half an hour, being the high water of this wave; and then as it began to recede, the surface of the La Plata was falling till 7.15 P.M. This wave is marked at its top by a large A, and we shall for the sake of convenience of expression call the wave for the future "wave A." By a similar proceeding we obtain the waves B, C, D, E, and F, all of which belong to the La Plata at Buenos Ayres; the lowest level between C and D waves being taken as the datum line.

Night and day are distinguished on the diagram by a shading of those hours which belong to night, namely, the space from 6 P.M. to 6 A.M. appears shaded, otherwise the numerous hour lines make it difficult to see which part of the day they belong to. The meridian passages of the moon (upper and lower) are also marked on the diagram at the exact hour and minute of transit of the meridian of Buenos Ayres, she being at the time close upon new moon. We have, moreover, shown the force and the direction of the wind, corresponding to every quarter hour of the day and night; this part of the diagram being a complete representation of the tidal phenomena of the La Plata from the 17th to the 21st January, 1871.

The rise and fall of the surface of the Paraná de las Palmas, as registered by the gauge at the section by hourly observations, is, for the same days and hours of day and night, shown on the diagram by another curve marked "Surface of the Paraná de las Palmas," the datum line of this curve being again the lowest level of the Palmas during the observations at the section. Referred to this datum, the gauge at the section at noon on the 17th January read $(2' \; 0'' - 9\frac{1}{2}'') = 14\frac{1}{2}$ inches, see page 62, which is the height of the Palmas surface above its datum for that hour of the day. On the night of the 17th, the Palmas gauge was not observed, the intention being to commence systematic observations on the 18th January at 6 A.M., the preceding day having been spent in preliminary and preparatory work for observations. By plotting the rise and fall of the surface of the Palmas on its datum as observed at the section for the hours of the 18th, 19th, and 20th January, we obtain by the periodical rise and fall a succession of waves, and we inserted at the top of each wave, commencing with the first on the evening of the 17th, the letters, little A, B, C, D, E, and F, for convenience of expression and reference to the diagram. We have accordingly on the Paraná de las Palmas a succession of six waves, and the position of these waves, their tops and their hollows, is absolute in reference to the horizontal distance from the tops and hollows of the La Plata waves, because they are fixed by the hours and minutes of the day, corresponding to the same instant of time at the gauge of Buenos Ayres and at the gauge of the Palmas section.

For the present we know nothing of the vertical distance between the tops and hollows of the Plata and Palmas waves, or the distance between their respective datum lines; at present we consider one datum line above the other, some distance apart, for convenient reference and comparison of the two systems of waves. Indeed, the distance between the two datum lines would for ever have remained a profound mystery, had it not been for an accidental circumstance which disclosed it.

A glance at the two systems of waves, the one of the La Plata, the other of the Paraná de las Palmas, will disclose the fact that there is a mutual dependence between the two. There is considerable irregularity in the size and in the level of the La Plata waves, owing chiefly to great disturbances arising from storms from various quarters; and we find, that on a reduced scale all the irregularities of the Plata waves are in succession reproduced on the Palmas; and considering that, besides the similarity of appearance, the waves of the Palmas are always very nearly the same distance in time following those of the Plata, the inference is very strong, if not conclusive, that the Palmas waves are but an extension of those of the Plata, and that consequently the Palmas at its head is subject to a tidal rise and fall, and that the tide of the Atlantic passes through the La Plata estuary and sweeps through the entire length of the Palmas, and enters the main river, the Paraná; where, on a continually decreasing scale, the wave propagates itself to an unknown distance.

The current observations on the Palmas, however, finally determine the point, confirming that the rise and fall of its surface is entirely of a tidal origin, and due to no other cause. The current trials midway the river at a permanent station disclosed the fact, that the current is continually changing; and, that there is an absolute dependence between the increase and decrease of the current and the fall and rise of the surface of the river at the same locality. The observations show that the current increases when the surface

K 2

falls, and decreases when the surface rises; and that the current is a much more delicate measure and test of the rise and fall of the surface than the gauge itself, which may be read to a quarter of an inch; the current observations, however, disclose a rise of less than a tenth part of an inch, entirely beyond the power of accuracy of any gauge. We have shown on the diagram another curve representing the velocity of currents during the observations marked "Velocity of current at Observatory O in feet per minute," and the outline of this curve reproduces the Palmas waves "reversed" on an exaggerated scale; the current curve ascends when the surface of the Palmas descends; and, vice versâ, the current curve descends when the surface ascends; and every undulation in the surface is much more marked in the current curve than in the outline of wave,—the greatest current occurring in the hollow of two succeeding waves, the smallest when the top of the wave is reached, and so on.

If the rise of the Palmas surface were due to an increase of volume which the main stream might pour into it at periodical intervals,—a supposition in itself highly improbable, because the immense volume of the Paraná could not, in the ordinary course of events, be sensibly increased or decreased several times every day,—it must have the opposite effect on the current; an increased volume and depth would necessarily increase the current at the same locality and not decrease it. The current observations, accordingly, finally determine the question that the rise and fall of the Palmas has entirely a tidal origin, and that the Palmas waves are but the waves of the Atlantic on a reduced scale.

The two systems of waves, as obtained for the La Plata and the Palmas by independent observations at each station, are most interesting. They are, in fact, but one system of waves, modified in outline, yet retaining their original form as they ascend the rivers. It is easy to see that, for example, wave B of the La Plata is wave B of the Palmas, C of La Plata corresponds to c of Palmas, and so on; because the waves of the La Plata happen to be of very different size, otherwise, if they were nearly similar on the La Plata, as it happens at new and full moon, it might be difficult or even impossible to say whether B of La Plata was little B or C or D of Palmas; for, the La Plata B, C, D, &c., would be like the Palmas, nearly similar in appearance, and we should have no distinguishing feature between them. Such a circumstance would not alter the conclusions we had come to, but we should not be able to trace the individual wave, and we could not tell whether it required six or twelve or more hours for the Plata wave to appear at the Palmas section.

It is matter for surprise that the little, almost insignificant, wave C of the La Plata, which on the preceding low water only rose 16 inches to make high water at Buenos Ayres, should have travelled 88 miles from Buenos Ayres, the greater part of which, viz. 64 miles, up the Palmas branch of the Paraná, reproducing itself on a reduced scale at the Palmas section; retaining the same horizontal length from the commencement of its appearance to its termination; the low water preceding little c being higher than the following one, precisely as it was the case with the La Plata wave C; the difference between the two being in their vertical dimensions only, which, by the time the C wave of the La Plata reached the Palmas section, was reduced to about one-fourth of its dimensions, appearing as little c of the Palmas. The scale of reduction of the vertical dimensions is, however, different with every wave, and seems to depend on the relative position or level on which the La Plata waves make their appearance, and also on their size; whilst, for example, wave E of the La

Plata reproduced itself at the Palmas section with more than one-half of its vertical dimensions, the others take various proportions between the limits of one-half and one-fourth of the vertical rise and fall.

We have already commented at some length on these La Plata waves when the tides of the estuary were under consideration, see page 26, and to avoid repetition we must refer to the description and the arguments given on the subject; the series of waves under consideration are remarkable, and were produced by the exceptional coincidence of various phenomena, such as new moon occurring at the time of her equinox and in her perigee; the sun being also nearest the earth, affecting the tidal wave jointly in the same direction as the moon; moreover, a succession of storms from a quarter by which the tidal wave on the La Plata is most affected—all tending to make the series under consideration of special interest.

The two systems of waves disclose the fact, that the tidal wave travelled from Buenos Ayres to the Palmas section in a space of time varying from $5\frac{1}{2}$ to $4\frac{1}{2}$ hours; the former rate corresponding to ordinary low-water level, the latter to exceptionally high water, and that at a mean level of the Plata, corresponding to the low water between waves D and E, the rate of propagation is exactly five hours for a distance of $87\frac{1}{2}$ English miles, or at an average rate of $17\frac{1}{2}$ English miles per hour. These facts are in conformity with the movement of free waves, their rate of movement depending on the depth of water within which they propagate themselves, and so it appears from these tidal waves under consideration; they reach the Palmas section faster when the depth of water is greater, the variation in time is therefore not an irregularity but in accordance with established laws defining the movement of free waves. The rate of propagation depending upon the depth, we may determine at what rate the wave at mean level reached the submerged mouth of the Palmas, and having found the time, we may determine at what rate it travelled within the Palmas, confined between banks. The average depth of the La Plata at mean water from Buenos Ayres to the mouth of the Palmas is about 10 feet; in this depth a free wave would travel at a rate of 12 miles an hour; the submerged mouth of the Palmas being 16 miles distant, it was reached in 1 hour 15 minutes; and, the remaining $71\frac{1}{2}$ miles were, within the Palmas, traversed in 3 hours 45 minutes, or at a rate of 19 miles per hour. The mean depth of the Palmas is 38 feet 9 inches, for a width of about 1165 feet, the sectional area of the Palmas being 45,200 square feet at low-water datum, corresponding to ordinary low water of the La Plata. In a depth of 38 feet, however, a free wave will travel at a rate of just 23 miles an hour; showing, that even in this wide river its propagation is not as fast as in the estuary, the resistance of the shallower parts on the sides not being overcome and balanced by the uniformity of greater depth midway the channel.

It is interesting to consider the true outline and form of a tidal wave. We have already observed that the waves, as shown on the diagram, appearing as a succession of abrupt upheavals of the surface, are enormously exaggerated; and that they lost every trace of resemblance to the true form of the natural wave, although for engineering purposes and reference the outline of every wave is correct. This arises from the difference in the horizontal and vertical scales of the diagram. The vertical scale is one-tenth full size, in other words, a rise of 10 feet of the surface of the La Plata is represented by 1 foot on the diagram. The scale of the horizontal distances of the diagram is, however, $\frac{3}{10}$ inch for every hour, which expression of time in reference to the wave represents the distance which it travels in one hour. We have just seen that the rate at which a wave travels depends on the depth of water within which it is propagated; and accordingly, the distance represented by one hour is different with every one of the waves, although in the present case the variation of distance is within about 20 per cent. We may, however, select one particular wave, and consider its true outline and dimensions. Taking for example wave E, remarkable in other respects, we may readily determine its dimensions, and the value of the hour on the diagram for this wave. By far the greater part of this wave is in deep water; for, two hours after its appearance at Buenos Ayres, it moved in water marking 7 feet on the gauge, and the measure of its propagation is the position of low water between E and F of La Plata, and E and F of the Palmas, which in both cases is sharply marked and equal to four hours, whilst the position of high water as to distance is somewhat vague on account of flatness of their tops.

Wave E was accordingly propagated at an average rate of about 22 miles per hour in the Palmas and the La Plata, the variation in depth being about the same in the estuary and in the river. The base of the tidal wave E measured by the gauge at Buenos Ayres from the time it appeared and disappeared, is 11 hours 45 minutes, corresponding to a distance of 258½ miles. The horizontal scale of the diagram for this wave is accordingly 110 miles to the inch of the scale, or $\frac{1}{6966960}$th part of the true size of the wave, and since the vertical scale is $\frac{1}{10}$th of the true dimensions, the exaggeration between the two scales amounts to 696,960, or in round figures to 700,000; that is to say, the vertical dimensions of the wave are drawn seven hundred thousand times larger than the horizontal distances. It is difficult to realize the magnitude of this exaggeration, by which every trace of similarity between the true form of the wave,—and as it appears on the diagram,—is effectually extinguished. If the line on the diagram marked "Ordinary low water at Buenos Ayres," were drawn so fine that it could just be perceived by the eye, all the tops of the waves, A, B, C, D, E, and F, would fall within the thickness of that line; the finest that could be drawn and seen. More than this: if by some means mathematical lines could be drawn and be seen through the most powerful microscope, magnifying a thousand times, the top of the highest wave, viz. that of F, would still appear less than $\frac{1}{100}$ inch from the straight line representing the datum. It follows, that the thickness of the finest line—and thickness it must have to be visible—is still a great exaggeration if one side of it be considered the datum and the other the top of the waves, but it would be a much nearer approach to their true form. The fact is, that the outline of the wave as it rises above the true level of the earth is exceedingly gentle, the opposite of what the diagram represents it to be; but if we attempted to draw its natural outline it would be impossible to represent it on any diagram otherwise than a straight line escaping all geometrical form and consideration. This will further appear if we consider the true form and dimensions of a tidal wave. The length of wave E is about 258 miles; the instant the wave made its appearance at the Palmas section it spread upon a level running through the 7-foot mark of the gauge at Buenos Ayres, through the entire length of the Palmas, along San Fernando, Buenos Ayres, Eusenada to Monte Video, and many miles beyond into the Atlantic. This is the size of one wave; and yet the top of that wave, at the time near Eusenada, was only about 30 inches above the level we had drawn through the gauge. It is important that the true form and the dimensions of tidal waves be understood; they should always be borne in mind, otherwise the phenomena attending the formation and the propagation of waves could not be conceived and comprehended.

It is necessary to make another observation in reference to the position of these waves. In the diagram we assumed the gauge of Buenos Ayres to be at the left margin, and the records commencing at 10 o'clock in the morning of the 17th were for every succeeding hour plotted from the left hand to the right, which produced the succession of waves called A, B, C, &c. The movement of "Time" was here assumed from the left to the right, and if we assume the same direction for the movement of the wave, then the Atlantic would be to the left hand and the Paraná to the right. It is, however, obvious that under this assumption we have to reverse each wave and also their succession; for, if the waves followed in that direction, from left to right, and the gauge had been stationary, wave B ought to follow A, and be to the left of it, and not to the right as in the diagram. Under that assumption the gauge ought to be drawn at the right margin of the diagram, next to it wave A preceding B, this again preceding C, and so on. As the diagram stands, with the gauge at the left margin, it would appear as if the waves proceeded from the Paraná towards the Atlantic, or the reverse of their true movement. It is usual to represent movement from the left to the right, and as the measure of movement in the diagram is the time elapsed, so the intervals representing time were plotted from left to right as their natural succession; we must then suppose the wave stationary, the gauge moving with the time, and to occupy in succession the position of the hour lines; the waves will then produce the outlines on the diagram. If, however, we assume the gauge stationary and the waves to move, which is the natural course of events, the case will be reversed, and so the diagram must also be reversed, and at the commencement of records all the waves must be to the left of the gauge.

We have shown on Diagram No. 2 wave E, as produced by the Buenos Ayres gauge; it changed outline as it ascended the Palmas, and the moment it reached the section its outline appears from the Palmas wave E, so that the wave is compounded from two diagrams; part of it within the La Plata being determined from the Plata gauge, and the other part within the Palmas from the Palmas gauge. This wave, as shown on Diagram 2, is imaginary; it never existed in that form, because it never proceeded in that direction; it was not the right-hand branch which entered the Palmas, but the branch to the left. Had the wave E proceeded from the Palmas to the Atlantic, it would have been its true outline the moment it left the section.

On Diagram No. 3 we have shown wave E reversed, as it ought to be if we assume the Atlantic to the left and the Paraná to the right, the waves proceeding from the Atlantic to the Paraná; their movement being then from the left to the right hand. We have shown the dotted position of the wave as it approached the gauge from the Atlantic at the instant it had reached it; the outline of the whole wave was then shown by the dotted line. As it proceeded farther and farther into the Palmas, its right branch, proceeding ahead, changed by degrees; and by the time it reached the section it was raised into the position shown by a hard line, and the other end of the wave was some distance beyond Monte Video in the Atlantic. The hard outline shows the entire wave E as it existed the instant it reached the Palmas section, viz. at 4 P.M., January 19th, 1871, Buenos Ayres mean time. It thus appears, that the waves change their outline as they proceed up the river, getting more and more reduced until their height becomes imperceptible; and, in consequence they seem to disappear altogether

72 THE PARANÁ DE LAS PALMAS.

## ANALYSIS OF INCLINATION.

We now enter on the consideration of a most delicate and difficult subject, namely, the analyses of the observations in reference to the inclination—or fall per mile—of the Paraná de las Palmas. All our attempts to get at some fact which might disclose it have been in vain, and at the completion of the survey of the Great Rivers we were as much in the dark on the subject as ever. It has already been mentioned, that the banks of the river are composed almost entirely of marshy land covered with a coarse very long grass, and that an accurate levelling operation, correct to within a fraction of an inch per mile, is impracticable. The easy and exact reproduction of the wave C of the La Plata—an insignificant elevation of its surface—for a distance of 88 miles up the river cannot fail to make the impression that the fall of the Palmas must be very small indeed; but what it may be there is nothing to show. We had on the main river, the "Paraná," where the tidal wave cannot be traced any more,—within a straight reach of the river, which on its left bank admitted the exact measurement and levelling of a distance of 4800 feet,—made a levelling operation with every possible care with a good 14-inch level adjusted, the staff being held on the tops of pegs equidistant, 300 feet apart, and the surface of the Paraná was smooth, with but gentle ripples; yet, after the completion of the operation—casting up the figures, the result of the levelling between two points of the river 4800 feet apart, was 0·00 foot; that is, practically level; and the conclusion we came to was, that either some error had crept in, or the fall is so small that it is beyond the power of the instrument to disclose it. By repeating the levelling several times, the difference of level between two datum pegs may be determined to almost any degree of accuracy by a good leveller and a first-class instrument; but the uncertainty arises from the unsteadiness of the surface of the Paraná, which appears more like a lake than a river. This unsteadiness may amount to an inch or more, and would perhaps be of not much account if the distance of the exact levelling could be extended to, say, 10 miles; where the fall obtained might be considerable, at any rate, ten times that of one mile approximately.

On the Palmas no such attempt at levelling could be made with any chance of success; on the "Paraná" the only chance that presented itself, small as it was, left the question open—vague; and at the conclusion of the surveys we inclined to the belief that, after all, some error must have crept in; perhaps the effect of excessive heat on the instrument, or some other cause, extinguishing the small fall which must have existed between the two points.

Two years after the completion of the surveys, engaged with the detailed consideration of the Palmas Observations for their publication, we placed the two systems of waves belonging to the Plata and to the Palmas one above the other; and, the difficulty arose how far the two datum lines should be placed apart; it was a vexed question, knowing nothing about the fall of the river. Ultimately we placed the two datums as far apart as the space would admit to clear one diagram of the other, and we proceeded to consider the waves in succession, and to analyze their effect on the currents. Each wave had its special features and its peculiar effect on the Palmas current. It was thought, that from the variation in the rise of the waves, their effect on the inclination might be traced by the effect on the currents, and the inclination itself might be found; but this proceeding always involved a theory—a certain dependence between current and inclination ; and if a theory was to be adopted, there was no difficulty to determine the fall from the currents observed at the section ; leaving us, however, always in doubt whether the theory was applicable for rivers so great ; and with the experience already gained by direct observation of currents, disclosing the defective basis on which our theories were based, there was little temptation to trust to results of calculations in no manner checked by observation.

We had completed the diagrams, the curve representing velocity of current, and also the analysis of the effect of the waves on the currents ; the observation of which terminated at 5.11 P.M., January 19th, under No. 44, at the permanent locality midway the river ; without, however, coming to any conclusion as to inclination ; and we were on the point to abandon the question, and about to enter upon another series of observations higher up the main river, when our attention was drawn to a " memorandum " in the Survey Book at the termination of observations on the Palmas section, and which will be found page 61, under " Memoranda on Currents, January 19, 1871," a literal extract from the Survey Book. The strange event there placed on record—a casual observation after the termination of the surveys and trials for the Palmas—could not fail to make a deep impression on any observer, and the more we considered the event the more remarkable it appeared. The current of the Palmas reversed at the section : What does it mean?

In the first place, certain it is that the tidal wave overpowered the river, which at the time did not exist as a river, for it had no current ; the little which could be traced was negative ; and, substantially the Palmas at the section was a lake and not a river. In the next place, certain it is that at the time the inclination of the Palmas was nil, slightly negative, for there was no current, and the little that could be traced was negative.

If, however, the inclination of the river was reversed at the section, however little it may be, it must have been reversed along the whole length of the Palmas, otherwise it could not have reached the section ; for, if at any point below the section the inclination had only once changed for a positive quantity, there was nothing to alter it, the tendency of the river being to maintain and rather to increase that positive inclination ; accordingly, that wave by overpowering the river produced a negative inclination along its whole course, and the river was running, however little, along its whole length up to the section in the opposite direction, until a point was reached where the wave and the river balanced, and where the former at last began to break the power of the latter, and to rise above it, and so at last create a positive inclination. The wave accordingly filled the Palmas, which below the section was unable to rise above it, and the whole of the Palmas was so filled and levelled by the wave to its very top. We must bear in mind, that below the section the river could nowhere have risen above the wave ; otherwise the negative inclination could not have made its appearance at the section.

The wave, then, levelled the river and converted it into a lake ; and by referring to the diagram it is easy to see that, on account of the long horizontal top of the wave extending over $2\frac{1}{4}$ hours, equal to a distance of about 55 miles, the greater part of the distance from Buenos Ayres to the Palmas section was under the power of the wave, and was levelled.

L.

converted into a sea of still water. This remarkable wave did then for us what we never could accomplish by the aid of our instruments; it made a true and perfect level between the gauge at Buenos Ayres and that at the Palmas section. A mighty wave came at last to assist us with a thousand millions of cubic yards of water to float us over our difficulties.

The memorandum in the Survey Book states that at 9 P.M., January 19th, there was a slight current from the Atlantic, about 8 feet per minute, just perceptible. If we refer to the diagram, we find wave little E of the Palmas to correspond to that hour of the day; and that the observation referred to was within one hour of high water occurring at 10 P.M., measuring 3 feet 9 inches on the gauge at the section; the Palmas rising in that hour a little more. The top of the La Plata wave E measured 9 feet 6 inches on the gauge at Buenos Ayres, the zeros of the two gauges would therefore be (9 feet 6 inches) − (3 feet 9 inches) = 5 feet 9 inches apart; and, since on the diagram gauge the datum of the Palmas was raised $9\frac{1}{2}$ inches so as to make it pass through the lowest water observed at the section, the difference between the two zeros on the diagram would be (5 feet 9 inches) + $9\frac{1}{2}$ inches = 6 feet $6\frac{1}{2}$ inches, through which line of the Buenos Ayres gauge the datum of the Palmas must pass. It happened, that the lowest level at the section produced by the lowest level of the La Plata, was ordinary low water of the estuary; we have accordingly the difference between ordinary low water of the La Plata and corresponding low water of the Palmas, $87\frac{1}{4}$ miles from Buenos Ayres, equal to $38\frac{1}{2}$ inches; or, at the corresponding ordinary low water of the two, a fall of 0·440 inch per mile. This figure for the fall of the Palmas is identical in the first two decimals with the fall of the La Plata, ascertained by a very different process, in which the elements for the determination of the inclination from ebb to flood approaching low water were the speed of the propagation of the wave and the fall of the surface in a corresponding time. Neither of these elements enter into the consideration of wave E; its speed of propagation is immaterial; the fall of the surface of the La Plata and the Palmas was nil, and yet the result is practically identical; and we may say the one confirms the other. We learn, moreover, the important fact that the Palmas and the La Plata have the same fall per mile, the one an extension of the other—the same thing in a different form; it is strong evidence confirmatory of the view expressed in the preceding pages, that once, ages gone by, the La Plata occupied the locality which the Palmas to-day holds, and that the great rivers are daily encroaching on the La Plata, and in time will claim it all as their own territory. The fall remains the same; as islands appear right and left, the channel between them deepens as imperceptibly as they are raised and built up from the river's deposit.

Wave E relieved us from a serious difficulty and uncertainty; it prepared us, however, at the same time a number of perplexities and apparent contradictions and a series of puzzles, which we must try and explain, and solve with patience and perseverance.

On the diagram the two systems of waves belonging to the Plata and the Palmas are shown in their true relative position, as determined by wave E, and the event which attended its progress up the river. The tops of Plata E and Palmas E being placed on a level line, all the other waves, A, B, C, D, and F, and the corresponding Palmas waves, A, B, C, D, and F, occupy their true relative position.

Let us first examine the difference between the various low-water levels of one system and the corresponding low waters of the other. Commencing with the low water preceding A on the La Plata and low water preceding a of the Palmas, we have a difference of 39 inches in the level of the two; in the same manner we have from the low waters preceding B, C, &c., of the Plata, and b, c, .... &c., of the Palmas, the following result:—

Difference between the low-water levels preceding

| | | | | |
|---|---|---|---|---|
| A, a, | .. | .. | .. | + 39 inches. |
| B, b, | .. | .. | .. | Not observed. |
| C, c, | .. | .. | .. | + 35½ inches. |
| D, d, | .. | .. | .. | + 38½ „ |
| E, e, | .. | .. | .. | + 23 „ |
| F, f, | .. | .. | .. | + 15½ „ |

All of which are positive, the low water preceding A on the Palmas is 39 inches higher than that preceding A on the Plata, and so on.

These figures show, that when the La Plata is near its ordinary low-water level, the difference between the two levels is nearly constant, viz. 38½ inches; that although the low water preceding C, c, is 3 inches short, it is accounted for by the powerful disturbance of the La Plata preceding that low water, the top of the B wave being excessive; and, although the run of water from the top of B was of unusual duration, the whole mass of the wave being suddenly released by the gale veering round from N.E. to north, yet, it was not long enough to allow the La Plata to fall to its ordinary low-water level, the tidal wave C having already made its appearance, checking any further fall. The figures also show that as the low-water level of the La Plata is raised more and more above its ordinary position, so the difference between the low-water levels of the Palmas and the Plata is more and more reduced. The wave E, by deepening the estuary 50 inches at low water between E and F above the ordinary level of the La Plata, also reduced the difference between the corresponding low water at the Palmas section to 15½ inches, being only $\frac{5}{12}$ths of the fall, or difference of levels, due to ordinary low water of the La Plata. So far we meet with nothing unusual or contradictory, which might be in opposition to accepted rules and principles.

When, however, we examine the relative position of the tops of the waves of the two series, we fall into a number of apparent contradictions and difficulties and puzzles, which make it appear as if the position of the two systems of waves, as determined by the E wave, could not be right. We find, that with the exception of the two small waves A and C of the Plata, the tops of all the others appear at a higher level than the tops of the corresponding Palmas waves. There would be nothing unusual in this circumstance were it not that these waves apparently contradict the inferences drawn from the E wave; for, if the position of the two systems be correct, and the top of E standing on a level with the top of little e had levelled the Palmas, checked and reversed its current, how much more effectually ought this to have been accomplished by the B, D, and F waves? the tops of which are considerably above the corresponding Palmas waves. Not only did these waves, however, not reverse the current of the Palmas, but there was at the minimum always a considerable current varying from 100 to 120 feet per minute down the Palmas towards the Atlantic. We should not have ventured to assign the relative position of the two systems of waves they occupy on the diagram, had that position been the result of an ordinary levelling operation, which for a distance of eighty

L 2

and odd miles might have been a couple of feet out of truth; considering, that such levelling involved hundreds of positions and thousands of figures, which may be affected more or less by as many more causes—temperature, refraction—in a country where mirages are of daily occurrence, &c. &c. We should under those circumstances have doubted the accuracy of our own levelling; but we doubt not the accuracy of nature's levelling, which, with one position, marked the level on both gauges by a true horizon independent of all the sources of errors to which we should have been subject. The position of the two systems of waves as determined by the E wave is absolute; we must patiently seek for causes and reasons to explain the propagation and generation of the other waves, and explain their relative position. We shall not attempt to follow the circuitous route which led us to certain conclusions, and which might fill a hundred pages with interesting arguments on hydraulic questions, but we will follow the shortest course, which may lead to the same result.

The reason why the B, D, and F wave did not level the Palmas and reverse its current, though apparently in a more favourable position to do so than the E wave, is in the first instance, because the E wave is the most powerful of all the waves, causing the greatest elevation of the surface of the La Plata above the low water at the Palmas section; both considered at the same instant of time. At 4.30 p.m., January 19th, the top of E wave was reached, and it measured 9 feet 6 inches on the gauge at Buenos Ayres. At the same instant of time, low water having just passed at the Palmas section, the level measured 7 feet 2 inches on the same gauge. The surface of the La Plata was therefore at that moment 28 inches higher than the surface of the Palmas at the section. None of the other waves, which at first sight appear in a more favourable position, could raise their tops anything like the same amount above the surface of the water at the section, considered simultaneously in the two localities. The B wave raised its top at 1 a.m., January 18th, 14 inches above the level of the Palmas surface; the D wave, at 4.30 a.m., January 19th, 14½ inches; the F wave, at 3.30 a.m., January 20th, 16 inches: so that the E wave rose about double as high above the surface of the Palmas at the section as any of the other waves; and, as far as the river is concerned, the rise of the wave above the river's surface is the paramount question—not the absolute height of a wave measured from a certain datum. A river is always trying to rise above an obstruction placed in its course, and the event proves, that if the Palmas was powerful enough to overcome waves at its mouth which appeared 14, 14½, and 16 inches respectively above its level at a point some 60 miles higher up, it was overpowered by a wave which rose 28 inches above that level; the wave pouring water into the river faster than the river could bring water down to meet it; so the river could not rise above the wave, and the wave ruled the channel of the river and reversed the direction of the current along the whole length of its course. It is clear that as far as the river is concerned wave E was by far the greatest and the most powerful of all the waves on the diagram, and that its effect on the river must have been the greatest.

In the next place, the strange circumstance that the less powerful waves should have overtopped their corresponding high-water waves at the section, viz. that the top of Plata B should be on a higher level than the top of Palmas B, and the same with D and F, may also be satisfactorily explained; and it will in another way show that these waves were only wanting the necessary power to raise the tops of the Palmas waves to the same level, and that their overtopping the Palmas waves is a proof of their weakness, and not of their strength, as it at first sight may appear.

The power of a wave consists in its height and in the volume of water which it carries with it. In the present case, taking for example the B wave, its power in reference to the Palmas River is measured by its height above that river at a certain point, which, in reference to the locality of the section, we found to be 14 inches; the volume which that wave may bring to bear on the river at the section is measured by the area enclosed between the low-water level at the section and the top of the wave. Now, this wave had to fill the channel of a river a quarter of a mile wide and sixty and odd miles long; and although it unquestionably did fill a considerable portion of that length, and probably reversed the current of the Palmas at its mouth, long before the section was reached the power of the wave was broken by the river, which at some point between its mouth and the section carried more water forward than the wave could pour in, and the surface of the river commenced rising parallel to its former position until the wave began to recede from the mouth of the river; and, as the wave could not bring water into the river without a negative inclination of the surface within the channel of the river, viz. a fall from the La Plata into the Palmas; so the level at which the two balanced some distance up the river must have been lower than the top of B wave at the mouth, as measured by the gauge at Buenos Ayres. The same argument applies to D and F waves; they were wanting in both, volume of water and in height, to level the Palmas as far up as the locality of the section.

Having removed the doubt which may have attached itself to the exact position of the two systems of waves of the Plata and Palmas, we may now proceed to consider the ever-changing inclination of the Palmas and the limits of variation of its inclination. We must, however, observe that we cannot attempt to deal with the mass of interesting features which the two systems of waves present to the consideration and attention of hydraulic engineers; we can but select a few leading facts for discussion and argument. The two systems contain most valuable records of facts, which should be carefully and minutely studied by those who care to acquire sound knowledge in this branch of the science of engineering. It is only by the aid of the "current curve" that we can feel our way along the complications of the two systems of waves; it is the "current" that we must constantly look to, and see whether our arguments keep in harmony with the laws of nature, and so prevent us from establishing imaginary laws; as easily conceived and defined, as evaded.

One of the striking features in the two systems of waves is, that the inclination of a river's surface at a given point is in no manner represented by the inclination or fall between that point of the surface and another some distance below or above it. In the examples we are going to refer to, the two points under consideration are no doubt a great distance apart; but this can only the more establish the rule; and by reducing the distance, it can only involve the question of degree. The La Plata, as far as Buenos Ayres, may be considered as an extension of the Palmas,—certainly so for our purpose,—both having the same inclination and propagating the tidal wave; but, if exception be taken to this assumption, the La Plata gauge may be supposed at the mouth of the Palmas; and it would have produced substantially the same system of waves, slightly reduced in height; the low-water levels would be about 6 inches higher, which would mark the 4-foot level on the gauge as the ordinary low-water level at the mouth of the Palmas; in every other respect the position of the waves would be similar, except about one hour nearer to each other; and the lines of the two systems would similarly rise and fall one above the other and intersect each other. To introduce, however, no alterations in the lines, it will be better to consider the gauges at their respective localities, 87½ miles apart.

We have very nearly the intersection of the B and D waves corresponding to their descent, viz. at 4.30 A.M., January 18th; but the current observations only commenced 1½ hour later; and, although we could determine the current at 4.30 A.M. by analogy—very near to what it was—we leave this intersection out of consideration, as it involves assumptions, and is not based exclusively on observation.

The D wave of the La Plata intersects C and D of the Palmas; the E wave intersects the D and E waves of the river; at all these intersections the fall between the two stations from surface to surface is nil; yet, at the first intersection occurring at 2 A.M., January 19th, we have the maximum current of the day—almost the maximum during the whole period of observations—at 200 feet per minute, and the current remained constant for about 2½ hours.

At the next intersection, happening at 7 A.M., January 19th, the fall of the surface between two points of the river 87½ miles distant being again nil, the current was 108 feet per minute. A line drawn from the surface of the water at one gauge to that of the other, would have been level.

At 1.20 P.M., January 19th, another intersection occurred, the fall being again nil between the two stations, the current was nevertheless 185 feet per minute. At 8 P.M. the same day, we have the last intersection during current observations; the fall between the two stations was again nil, but the current was not only nil, it was negative, reversed, about 8 feet per minute.

From these few examples, the inference might be almost safely drawn that the "effective" inclination of a river's surface at a station may have nothing in common with the inclination obtained for that locality by drawing a line touching the river's surface at the station, and at another point a considerable distance below or above it. The inclination so obtained may, in contradistinction, be called the "imaginary" inclination of the river at the locality under consideration.

The above examples do not, however, by any means exhaust the strange events placed on record by the observations; exposing the weak points of our theories. We have just considered the "intersections" of the La Plata and the Palmas system of waves, which were as many levels drawn between the two stations. But let us follow and consider what happens past those intersections. The line of D wave of the La Plata having intersected the line of the C wave of the Palmas—the former continues to rise above the surface of the Palmas for hours—it remains above it for five hours, and during all this time we have a powerful current, never less than 125 feet per minute towards the La Plata, and uphill; if the inclination of the river's surface between two localities of its channel were to be the measure of the inclination at a certain point under consideration. The absurdity of the proportion is still more heightened if we further consider that for $2\frac{1}{4}$ hours after the intersection, when the maximum uphill incline is reached, the current remains constant—200 feet per minute invariable—taking not the slightest notice of the elevation of its surface at its mouth above the level where the currents were observed; and, at that time wave E had already considerably entered the confined channel of the Palmas.

The same thing happened on a larger scale with the E wave of the La Plata. This wave, after its intersection with little D, rose for many hours above the Palmas level. Nearly the maximum current was reached as the wave, at 1.20 P.M., January 19th, rose above the horizon of the Palmas level, and as it continued to rise above it for three hours, the current was all the time increasing, reaching its maximum within 3 inches of the top of the wave, standing at the time 26 inches above it, with a current of 192 feet per minute towards the La Plata, or "uphill." Wave E broke over the whole length of the Palmas and reversed its current, and making confusion worse confounded at 9 P.M., when the surface of the La Plata was 11 inches below that of the Palmas, or a positive fall from the Palmas into the La Plata, the current at the section run in the opposite direction from the Plata into the main river above the section.

If the Palmas always insisted to run "uphill" for hours and hours together, we might reasonably suspect that there must be something wrong somewhere, either in the distance selected, or something peculiar to the locality, or due to tidal effects not explained, &c. &c.; but we will presently show that the same mode of determining the fall, or inclination, of a river's surface at a given locality, will also give correct results with the Palmas, and they are correct in spite of distance, or tidal influence, or exceptional locality; and this circumstance is the most powerful argument against the usual and accepted system of proceeding to determine the fall of a river in a given locality by the fall between two distant points of the river's surface enclosing that locality. The Palmas is nothing but an extreme case, where the errors and fallacies of the system usually adopted are very marked, and are prominently shown. That system may give the true fall, or inclination, at the locality by chance, but in nine cases out of ten it will be erroneous; and, considerably so.

Taking for example the fall between the surface of the Palmas at the section, and of the La Plata about two hours before the latter reaches its low-water level, we come at once near the true fall then existing with every one of the waves. We meet no more glaring contradictions; and the rule applied at that hour does not only give us results in harmony with the observations and established and tried principles of hydraulics, but happens to bring us near the mark in each case. Instead of guessing, however, the result by the application of a defective rule at a favourable moment, let us at once ascertain it by a systematic proceeding.

The inherent defect of the rule is, that it assumes the inclination of the river's surface, between the two distant points selected for the purpose, to be a constant quantity; that it assumes the fall to be throughout the same. There are but few rivers, and fewer localities along their course where this assumption may be near the truth; in most cases it will be far from it. The Palmas is a river of a uniform channel in width, depth, and sectional area, running through a level marshy land without any obstructions in its course, without any tributaries of significance; if this river were allowed to adjust its fall to the volume it conveys from the main river to the Plata, there can be no doubt but that its fall would be uniform throughout. The great disturbing cause of the river is the rise of the tidal wave on the La Plata, and although its fall is also a disturbing cause, the former is by far the greater, acting against the river, whilst the latter operates with it, in the same sense and direction. The Palmas soon adjusts its inclination in accordance with the receding tidal wave; and, having so adjusted it, its surface falls nearly parallel to its former position throughout its whole length with the fall of the surface of the La Plata. This is clearly shown by the current curve, which rapidly increases at the change of an exceptional high water of the La Plata, the current assuming its normal velocity corresponding to the ordinary inclination of the Palmas. The B wave was an exceptional rise on the La Plata; and at one time it reduced the current at the section to nearly one-half; within about four hours the current, however, re-established itself, and, with little ups-and-downs, remained constant; averaging about 200 feet per minute for eighteen hours. During these eighteen hours the rise and fall of another tidal wave C of small dimensions is included, of which the current of the Palmas took but little notice, rising easily above the wave; and although it did affect the current in the exact outline of the little Palmas wave c shown reversed by the current curve, yet the maximum reduction was not one-tenth of the normal current, and the effect on the inclination was considerably less.

The ordinary rule may be applied when there is evidence that the surface of the river assumed a permanent and uniform fall between the two stations at which its level had been observed and ascertained. This happens on the Palmas just about the time of low water of the Plata, when the Palmas and the Plata had been both running for many hours together in the same direction; but we must bear in mind, that the full effect of the low-water level of the Plata can only be felt at the Palmas section about four hours later—the time of its propagation; and, if the low water of the Plata remained permanent, so would the corresponding low water at the section, and then the inclination of the Palmas surface would remain uniform throughout. The Plata low water does not, however, remain permanent, but after about an hour's delay, begins again to rise—at any rate, for an hour or two the change in its low-water level is small, not material, and consequently this change could only be felt an hour or so after the low-water level on the Plata had its full effect at the section. If we therefore measure the distance between low water on the Plata and corresponding low water on the Palmas, and connect the two by a line, we obtain the true inclination of the river's surface which was "effective" along its whole length for about an hour, or for about 20 miles' stretch at a time; and would have remained permanently so along the whole length, had the Plata remained at that level. It is, however, immaterial for our purpose whether that line drawn between the two low-water levels runs parallel to the inclination of the Palmas along its whole length or not; we are content, and indeed do not want more, than that the line be parallel to the tangent of the inclination at the locality of the observed current; nothing more is wanted, and we do obtain it with accuracy by the above proceeding.

These considerations also disclose the error of measuring the level at each station at the same instant of time; and in a great measure it is due to that circumstance that the absurd and contradictory results were obtained when the inclination of the waves intersecting each other on the two systems was analyzed.

For the tops of the waves we cannot determine from their position the "effective" inclination of the river's surface, because we do not know how much the waves lost from their tops as they entered the Palmas and poured their water into its channel. With the exception of E wave, the top of which reached with undiminished height the Palmas section, all the others were reduced long before they reached the section, and as we do not know how much, having no intermediate stations on the river, we cannot, from the outline of the waves, determine the fall at the section corresponding to those hours of the day; but we may, by the aid of the current curve and the "effective" fall at certain hours, also find those inclinations by calculation.

The difference of level between the two stations, 87½ miles distant, and the fall per mile at the various low-water levels, with the corresponding superficial currents midway the river in a depth of water varying from 49 feet 4 inches to 50 feet 1 inch for the low waters for which the currents were observed, will appear from the following statement :—

| No. | Date of Low Water of La Plata, January, 1871. | Difference of Level between Low Waters of Palmas Section and La Plata, Time of Palmas Low Water. | Superficial Current in feet per minute at Observatory midway the Palmas. | Depth at Observatory. | | Fall per Mile. |
|---|---|---|---|---|---|---|
| | | | | ft. | in. | inch. |
| No. 1 | 17th Jan., 11 A.M. | 39 inches, 4.30 P.M., Jan. 17th | 213 | 50 | 1 | 0·446 |
| ,, 2 | 18th Jan., 12.30 P.M. | 36 ,, 6 P.M., Jan. 18th | 207 | 49 | 7 | 0·411 |
| ,, 3 | 18th Jan., 10.30 P.M. | 38½ ,, 2.30 A.M., Jan. 19th | 201 | 49 | 4 | 0·440 |
| ,, 4 | 19th Jan., 11 A.M. | 23 ,, 4 P.M., Jan. 19th | 193 | 49 | 11 | — |

Every wave on the Plata is a cause of disturbance to the inclination and currents of the Palmas, and when these disturbances are regular and periodical, the river settles down and acquires its normal declivity. When, however, the disturbances are irregular, it is difficult to trace the inclination the river may have acquired at the time of such disturbances, even with the aid of numerous diagrams on levels and currents. The current, indeed, is the only reliable guide, and alone sufficient at once to determine the question, provided we are acquainted with the law existing between current, depth, and inclination; for, at a given locality and a given depth the current is entirely ruled by the inclination. We are, however, now engaged to trace the inclination of a river's surface by the difference of level between two distant points along its course.

The inclinations corresponding to No. 1 and No. 3 low waters, are obtained for ordinary and regular waves falling to near ordinary low water, and no serious disturbance preceded them. We find that the river had time to settle down, and that a line drawn between the levels of the two stations truly represents the inclination along the whole river and also at the section, and that the currents observed are in harmony with those inclinations. The line drawn between No. 2 low waters falls already a little short of the true inclination, the preceding B wave being exceptional, and although followed by an exceptionally long run, the time was not enough to make the difference of levels so adjust themselves that the rule applied could have disclosed the exact effective inclination. No. 4 is an instance

M

where a powerful disturbance affected the surface of the La Plata, its low-water level being raised considerably above its ordinary position; and a much greater disturbance follows it. Here the rule evidently begins to fail considerably; the current adjusted itself over the top of Palmas D in an identical manner as over Palmas B, the maximum current falling only about 10 feet per minute short, and the effective inclination was in both cases nearly the same; yet the rule fails to disclose it, because the approaching storm prevented the La Plata falling to its ordinary level, or at least near to it, thereby vitiating the rule which could no more disclose the effective normal inclination which the river rapidly assumed; shown by the currents observed at the station.

We have dwelt at some length on the question of inclination of the surface of the Palmas, analyzing the numerous observations which refer to it directly or indirectly. We believe our knowledge on the subject of inclination of a river's surface to be unsatisfactory. We often hear and read of a river's inclination to be so much per mile, which is about as correct and as descriptive as if a railway engineer said the inclination of his line is 1 inch per mile, because the difference of level between his terminal stations, 100 miles apart, amounts to just 100 inches. Do not railway engineers determine the inclination of their line at every change, and express it at so much per mile, or rather more scientifically, as one in a hundred, or in two hundred, or in a thousand, as the case may be? And has it occurred to many hydraulic engineers to follow a similar process along the course of their rivers? And, if on a railway between London and Liverpool, the substantial variations in the inclination may number ten, the substantial variations of an ordinary river's inclinations within a similar distance would number hundreds, and to "mean" all these inclinations of the river is an error more grave and strange than that of a railway engineer would be, who insisted on the above grounds to say, that the inclination of his line was 1 inch per mile, and that he cared not for local variations. But what are we to say of those who speak and write of the inclination of a river's bottom? The undulations of its surface are numerous and troublesome enough, and these govern the velocity of its currents; they are often too delicate for our best instruments, and sometimes escape our most refined means to ascertain them; and then to speak of a rough irregular bottom, varying inches and sometimes feet in a couple of yards' distance, as a measure of the river's fall or inclination, —it is too erroneous, and the sooner it is forgotten the sooner hydraulic science will stand on a level with the more favoured branches of engineering.

It so happens that rivers sometimes maintain for great distances a certain depth with but little variation, and thus the general inclination of the bottom will be nearly parallel to a line drawn to "represent" the surface; and the error of the proceeding being already grave, it matters not which line we take, the surface line or the bottom line, the result will be about the same, equally "correct." Although the results so obtained may be similar, the proceeding is nevertheless radically wrong.

It may perhaps be thought, that the Palmas is not a fair representative of rivers generally, that it has a very gentle inclination, and is, moreover, affected by tides. We do not see the force of the argument. Many rivers are subject to tidal influence, and their importance is usually enhanced by the circumstance of being subject to tidal phenomena, because they form then a direct communication with the sea, and their currents become of paramount importance to maintain good communication, or to establish it. The inclination, moreover, only affects the distance to which tidal range extends within the channel of a river, but in no way the laws and the phenomena we had endeavoured to trace on the Palmas. Nor is it correct to suppose, that rivers free from tidal influence are not subject to similar disturbances and variations in their inclination and currents, as the Palmas observations had placed on record. A short tributary suddenly flooded by a heavy fall of rain in a mountainous district, has precisely the same effect on the main river as a tidal wave; the only difference being that the effect of the tributary is irregular, and not periodical. Moreover, every irregularity in a river's channel is an obstruction, only to be overcome by a specific inclination at the locality where the irregularity may occur. Changes in the sectional area as well as changes in the depth of a river constantly affect its inclination, which cannot be identical in two localities along its whole course, where we may have similar sectional areas, because the depths will be different producing these similar areas, and it is only by the similarity of their combined effect that similar inclinations may be found.

In artificial channels, free from external disturbances, having not only similar sectional areas but also similar geometrical form, the case is different; with these the inclination will be a constant, unchangeable quantity; and the solution of problems comparatively very simple. There is as much difference between an artificial channel and the natural bed of a river as there is between the equation of a straight line and that of a sinuous transcendent curve; the one represents simplicity, the other complicity. And this will also explain why we possess so little substantial knowledge of rivers. Nearly all experiments and observations had been conducted on artificial channels, usually of small dimensions, and the rules derived from those observations were freely applied to rivers; a proceeding about as correct as if the properties of straight lines expressed by equations were to be applied to the conic sections or to curves of a higher order. In an artificial channel the fall between any two points of its surface will nearly always represent the true inclination; and the fall of the bottom of the channel will also be nearly the inclination ruling the channel; and this may explain the origin of the erroneous rules applied to rivers, which are " meaned " in every imaginable way; we have mean depth, mean superficial currents, mean currents at section, mean inclinations; but we never can deal with a single depth, or a single current of any kind, or with the single effective inclination at the locality under consideration.

## ANALYSIS OF CURRENTS.

The current observations on the Palmas are exceptional. It is usual to observe the currents on the line of section at or below the surface, to ascertain the volume passing the locality in a given time, or to determine the force of the current at particular points across the river. It is, however, not usual to observe the current of a river at the same locality midway its channel every hour, and even fractions of the hour, day and night without intermission for days together. This system of observations originated with a suspicion, that the tide might affect the currents at the locality selected as the most suitable to ascertain the volume of the Palmas. Soon after our arrival, and the fixing of the gauge at the river's margin, it was found that the surface of the Palmas was slowly but uniformly falling, which in itself did not awake suspicion; but when after about five hours' fall it commenced rising again, there was room for considerable doubt on the point; we were, however, determined

M 2

not to abandon the locality without a thorough trial; for it may be that, even with the tide affecting the currents, they might be periodical, and then the volume may nevertheless be accurately ascertained.

The current observations on the Palmas consist, therefore, first, of a series of systematic trials of the river's currents at a permanent station, moored about midway the river, and called the Observatory; and second, of another series of observations at different points on the line of section across the river. The former, to ascertain the change of current, if any, at the same locality of the river; the latter, to ascertain the various currents of the river corresponding to various points on the line of section. There had been, moreover, trials at the Observatory to ascertain the mean current of the river by the current integrator, and also to determine the current near the bottom of its channel.

The results of the various observations are represented on Plate IV. by the several diagrams illustrating the nature of the currents. On Diagram No. 1 the rise and fall of the surface of the Palmas at the corresponding hours of the day is marked "Surface of Paraná de las Palmas." The superficial currents at the Observatory for the hours of the day are shown by another line, marked "Current at Observatory in feet per minute." These observations commenced January 18th, 6 A.M., and terminated January 19th, 5 P.M., with an additional observation at 9 P.M. The scale of the current curve is 30 feet to the inch; that is, the distance passed over by the current in one minute is plotted on the diagram accordingly; or, $\frac{1}{360}$th part of the full distance run by the current in one minute.

From the commencement of the observations at 6 A.M., January 18th, there was a rapid increase in the velocity of the surface current at the Observatory. For the first three hours the current increased 20 feet per minute every hour, increasing 50 per cent. in three hours on its velocity at 6 A.M. The rate of increase was then reduced, averaging only 6 feet per minute for the next three hours, and at half-past twelve the surface current reached 200 feet per minute, being an increase of about 70 per cent. since 6 o'clock in the morning. For the next sixteen hours the current was oscillating above and below the 200 feet per minute line, changing but little, and sometimes for hours the change in the velocity of the current did not amount to 1 per cent.; sometimes for a couple of hours the variation was nil, and within the margin of the possible accuracy of most careful observations; the current, in fact, was constant. Nor are the oscillations a succession of irregular ups and downs; the curve records a systematic increase or decrease for several hours in succession; and, it is evident, that these slight alterations are not produced by inaccuracies of the individual observations; otherwise, if they were not due to slight changes in the current, the variations, if due to inaccuracies of the observations, would in every succeeding hour appear either a little in excess or a little deficient; in short, once appear positive, and next time negative; but they could not remain positive for several hours and negative for many hours in succession. Indeed, we have ample evidence that with but few exceptions the currents as represented by the curve are correct within the thickness of the line.

Looking at the outline of the current curve for the sixteen hours following the noon of the 18th in a general way, the current appears to oscillate round the 200-feet line, the maximum variation being about 6 per cent. from the mean surface current during that time, and for all ordinary purposes it may be considered a uniform current of 200 feet per minute.

From 4.30 A.M., January 19th, the velocity of the current decreased rapidly, and within the following three hours the current was reduced to nearly one-half, at a rate of about 31 feet per minute; decreasing 93 feet within the three hours. This fall is more abrupt than the rise at the same hours of the preceding day.

From 7.30 A.M. to 9 A.M., January 19th, the fall was not only stopped, but the current remained remarkably constant; for one hour the variation being nil.

A sharp rise then followed from 10 A.M., January 19th, amounting to 35 feet in the first hour, and at a rate of 16 feet in each of the succeeding two hours; the current then continued to oscillate for about three hours round the 190 feet per minute line, when at 4 P.M. it again commenced descending rapidly, and especially after 5 P.M., being in the next few hours reduced to nil; and at 9 P.M., January 19th, it was found reversed, about 8 feet per minute from the La Plata.

During the 18th and 19th January the variations in the velocity of the superficial current at the same locality were so great and irregular, that at the time of the survey we abandoned the idea to determine the volume of the Palmas discharged into the La Plata by observations at the locality of the section. There appeared, at the time, no difficulty to ascertain the volume of the Palmas on a particular day, for which systematic observations had been made; but there was nothing to show what the volume might be the next day without repeating all the observations; and our time was limited for the purpose. We thought the observations might perhaps be of some value to the science of engineering, not suspecting, however, the treasures which lay hidden in the records of those trials.

Comparing the line representing velocity of currents at the Observatory, and the line representing the position of the surface of the Palmas, we see at a glance that there is a dependence between the two; that, generally speaking, the ups and downs of the one appear for the same hours of the day reversed in the other; that while the surface of the Palmas was falling the current was increasing; while the surface was rising the current was decreasing. The minimum current corresponds to nearly the top of the Palmas wave, in time about one hour preceding high water; the maximum current corresponds to low water on the Palmas about one hour in time preceding the lowest level. The current seems to be extremely "touchy" at the change of movement of the Palmas surface. Whether the change be for the rise or for the fall, it is most affected in the first couple of hours; a fall of 1½ inch per hour in the morning of the 18th increased the current 20 feet per minute each hour for the first three hours; a fall of 1½ inch per hour for the first two hours on the 19th, viz. from 10 to 12 A.M., increased the current 26 feet per hour; whilst a fall at a rate of 1 inch in another level, from 10 to 12 P.M., January 18th, hardly had any effect on the current, increasing it only at a rate of 3 feet per minute. A rise of 3½ inches per hour for the first two hours, from 5 to 7 A.M., January 19th, decreased the current 31 feet for each hour;

whilst a rise of ½ inch per hour, from 7 to 10 P.M. on the 18th, only decreased the current 6 feet for each hour; a sudden rise of 12 inches per hour, from 5 to 6 P.M. on the 19th, however, nearly annihilated a current of 170 feet per minute.

It will be seen from these figures, that the rate of fall of the Palmas surface is nearly the same for each wave, and the current having reached its ordinary velocity the continuation of the same fall does not further affect it, and we do not think that the slight variations in the rate of fall in the various waves, amounting to about a quarter inch more or less, accounts for the great variation in the currents; which in neither case will be due to the rate of fall, but to the circumstance that at the turn of tide the river takes about three hours to regain and to re-establish its ordinary inclination due to the volume it has to convey from the main river; and, having established it, the surface continues to fall nearly parallel to its former position in harmony with the fall of the surface of the La Plata; the movements and the currents of the two being then parallel to each other.

It is different, however, with the rise of the Palmas surface. The rise being due to tidal effect, as ascertained in the preceding pages, it is in opposition to the movement of the river, and its effect on the current seems to bear a definite proportion to the amount of rise. Thus we find that in the small and large rises the value of 1 inch rise per hour corresponds to a reduction of the current of close upon 9 feet per minute; and, had the observations been extended over many days the effect on the current due to a certain rise might have been defined; and from the rise and fall of the surface of the Palmas the corresponding currents might have been approximately determined, near enough for ordinary purposes, without resorting to laborious current observations.

We have now considered the first series of observations properly belonging to the series, all currents being observed for superficial velocity, at the same locality midway the river and in the same manner, between two boats 12 feet apart supporting a platform; the centre of the meter being held about 12 inches below the surface. The observations, extending over day and night for every hour,—and frequently to fractions of an hour without intermission for two days in the middle of the summer of an exceedingly hot climate,—could not all be made exclusively by ourselves. We had the valuable assistance of three Argentine engineers of good education and ability, who occasionally relieved us to allow the necessary amount of rest. Señor Don Cárlos Olivera conducted the observations Nos. 9, 10, 12, 14, 31, 32, 33, 34, 35. Señor Don Guillermo White, those of Nos. 15, 16, 17, 18, 19, 20. Señor Don Z. Tapia, Nos. 21, 22, 23, 24; in all nineteen observations were conducted and registered by our assistants. All other observations of these surveys were our own.

Some interpolations had been made in the first series of special importance and significance. We had at the same locality made trials referring to the surface velocity of current and a short distance below the surface, with a view to ascertain where the maximum velocity would be found. The trials appear under No. 11, and were conducted with the utmost attention, knowing beforehand that the difference, whichever way it would establish itself, could only be small; the difference of immersion being only 3 feet, and was called "Surface velocity and velocity 3 feet below surface," although the currents actually measured are 6 inches lower in both cases to admit the meter being covered at the surface trial, the diameter of its screw being 6 inches. These were "two-minute" trials by wire, and the velocities found may be relied on as correct at least in the first decimal. The true difference in the velocity of the two currents is not, however, that shown by subtracting one from the other, which would be correct if the current at the time had remained constant; the observations show that the current of the Palmas had been slowly and uniformly decreasing, and that even in the short time within which the two trials followed each other there was a slight change in the surface velocity, a reduction of 1 inch per minute, as shown by the curve on a large scale at Detail A, Plate IV. These trials were followed by a number of "minute" observations in rapid succession, confirming the result of the main trial, that the maximum current lay at the surface, and that for a current then in force, and a depth of 50 feet at the locality of observation, the current decreased just 1 foot and 2 inches per minute for every foot increased distance from surface to bottom within the limits of the trial. The weather was a perfect calm, and the heat oppressive; there was no visible disturbance of any kind, and the effect of the tide had been eliminated in the diagram.

The result of these trials is of some interest, because it seems to be doubted whether the maximum current is at the surface or some distance below it. It is unquestionable that, in a regular channel free from local disturbances, the maximum current establishes itself at the maximum distance from the bottom, in short, at the maximum distance from the retarding force; and it must be obviously and necessarily so. Gravitation moves every particle of water within the channel of a river; here, the force of gravitation is a constant quantity, affecting every particle of water in the same manner. The free movement of the particles is opposed by the resistance of the channel, a force generated by friction. The farther from the resisting force the particle of water may be, the less will be its influence felt, and the more will gravitation assert its power, and the more freely can the particle follow the latter force. So far there can be no doubt the argument will hold good. It is another question what will be the law of decrease or increase; we may propose a theory, but we ought to be guided by experiment and observation alone. As soon as we introduce a new opposing force the case will be altered; if we do not, however, happen to know either the power of the new force or its locality, it is idle to consider its effects. The resistance of the air may occasionally be considerable. The friction between air and water is illustrated in an accumulated form by the powerful waves which appear on the even surface of the ocean within one hour after a local storm made its appearance; it is also illustrated by the tidal diagrams, where a storm raises the mean level of the whole Plata estuary above the mean sea-level, the incline of which balances the friction of the storm on the surface of the water. So, mathematically considered, there is a resistance or friction between a calm atmosphere and the surface of a river, but this resistance is exceedingly slight, disappears and escapes our most delicate means to measure it; and it is idle to consider the effect of a grain in the scale, while we are still uncertain as to the number of tons which may fall into either scale of the balance.

The effect of a storm on a river's surface is not insignificant, and will sensibly reduce or accelerate the surface currents, according to the direction of the storm and the current. These matters are of scientific interest although of no practical value, because the opposing or favouring forces of winds are extremely variable, independent of all the elements which determine the movement of rivers within their channels.

Another interpolation made in the first series of observations, and of greater importance, refers to trials of mean currents at the Observatory, and to the velocity of current near the bottom of the channel at the same locality. These trials appear under Nos. 42 and 43, and which we now propose to analyze. The mean current of the Palmas at the Observatory midway the river corresponding to 4h. 28m. 49s. P.M., January 19th, as determined by the current integrator, was 159 feet 5 inches per minute. The surface current taken immediately after the "mean current" trial, viz. as soon as practicable, allowing a few minutes' rest to the men lowering and raising in succession the whole apparatus to and from a depth of 50 feet, and corresponding to 4h. 33m. 30s. P.M.,—was 184 feet 6 inches per minute ; the preceding mean, and the surface current under consideration, being "five-minute" observations, the result is accurate to the inch of velocity. These trials were promptly followed by an observation of the current near the bottom of the channel, which at 4h. 36m. 30s. P.M. was found to be ·129 feet 4 inches per minute, 49 feet from the surface and 1 foot from the bottom of the channel. The values of these figures cannot immediately be compared to one another without some deviation from the truth, because the velocity of the Palmas currents was during the time of these observations constantly decreasing; and, although all these trials from their commencement to their termination only occupied 10 minutes in time, there was, according to the velocity curve of the diagram, a decrease of about 4 feet in the velocity of the surface current during that time.

The diagram called "Detail at B," however, by the aid of the surface velocity curve, at once determines what the exact surface current had been corresponding to the instant of time for which the mean or the bottom currents were obtained. Thus, we find by the diagram, that the surface current corresponding to the exact time for which the mean current of 159 feet 5 inches was obtained, is 186 feet 6 inches per minute; and that the mean current at that locality was 85·46 per cent. of the surface current. If we now ask what would be the bottom current $x$ at that time, which united by a straight line with the observed surface current would produce a mean current equal to that ascertained by the integrator, we have the simple equation :—

$$\frac{\text{Surface current} + \text{Bottom current}}{2} = \text{Mean current}\;;\;\text{or,}\;\frac{186·5 + x}{2} = 159·4,$$

from which $x = 132·3$ feet per minute as the required bottom current. This specific current was not observed, because at that time we were integrating the currents; but we have an observation of the bottom current within $7\frac{1}{2}$ minutes from the calculated one, which was found to be 129 feet 4 inches. The surface current corresponding to the observed bottom current was, from the diagram, 183 feet 4 inches. In the space of $7\frac{1}{2}$ minutes the surface current, therefore, fell from 186 feet 5 inches to 183 feet 3 inches, or was reduced by 3 feet 1 inch; whilst the bottom currents were reduced from the calculated velocity of 132 feet 4 inches to 129 feet 4 inches, or 3 feet 0 inch, an agreement which could hardly be expected to come nearer, and is remarkably close ; considering, that the curve may deviate 1 inch in velocity either way, a deviation hardly visible on the diagram.

Again, having the surface current of 183 feet 4 inches corresponding to the observed bottom current of 129 feet 4 inches, we can determine what the mean current would be had it been possible to measure it simultaneously with the other currents; and, if a straight line uniting the surface and bottom currents would represent the actual or the mean of all the various currents from surface to bottom, the mean current would have been, under these circumstances, $\dfrac{(183 \text{ feet } 4 \text{ inches}) + (129 \text{ feet } 4 \text{ inches})}{2} = 156$ feet 4 inches per minute, and if this had been the true mean current, it would have been 85·27 per cent. of the superficial current determined from the diagram. If we now ask what was the percentage of the observed mean current in reference to the then existing surface current which preceded the latter surface current $7\frac{1}{2}$ minutes, we find it to have been 85·43 per cent., an agreement which seems to determine the question that a straight line uniting the bottom and surface currents is a true mean line, if it should not at the same time also represent the law of decrease of all the currents from surface to bottom; certain it is, that the line will so intersect the current curve, whatever it may be, that the currents are equalized thereby; the straight line in effect representing them. In the preceding chapter on the La Plata currents, we have already seen, that similar results were obtained by uniting the surface and bottom currents. The trials on the Plata confirm those on the Palmas; the Plata observations, however, also determine the straight line uniting the two, to be not only a true mean line, but also to represent the currents from surface to bottom. We have accordingly on the Palmas, within 10 minutes, a series of observations of consequence to river engineering.

Another mean current observation followed the preceding one within 20 minutes, the object at the time being to check the first, nothing else. The mean current of the Palmas obtained by the current integrator at the same locality, and corresponding to 4h. 48m. 28s. P.M., was 137 feet 6 inches per minute. A surface current observation immediately followed, and corresponding to 4h. 52m. 30s. P.M., gave a current of 163 feet 5 inches per minute. We cannot again immediately compare these two results; for, the diagram shows, that the current had been decreasing much faster than in the preceding mean trial, about three times as fast; that the mean current as obtained by the integrator for 4h. 52m. 30s. P.M. was corresponding to a surface current then in force equal to 170 feet 4 inches per minute. In this case the mean current is 80·71 per cent. of the surface current, and we can now, from the surface and mean current, determine the bottom current then in force by a rule established in former observations of a similar nature, which, accordingly by calculation, was 104 feet 8 inches per minute.

If we now compare the results of the two mean current observations, it appears that whilst the surface current decreased from 186 feet 6 inches to 170 feet 4 inches the mean current expressed as a percentage of the surface currents, decreased from 85·43 to 80·71 per cent.; that whilst the decrease in the surface current from the one mean trial to the other was 16 feet 2 inches, the bottom currents decreased from 132 feet 4 inches to 104 feet 8 inches, or by 27 feet 8 inches, all of which shows, first, that at the same locality the percentage expressing the mean current by the surface current changes with a change in the surface current; and that, accordingly, the mean current cannot be represented at a fixed percentage of the surface current, not only not in the same river, but not even at the same locality; second, that to a certain change in the surface current the corresponding change

N

of the bottom currents is much greater at the same locality. What is more, the current observations on the La Plata are in harmony with these several conclusions, which circumstance greatly enhances their weight. In the Palmas we were in a magnificent river of uniform section, 50 feet deep. On the La Plata we were in an open sea, with a uniform depth of about 25 feet for miles; yet, the results derived from the observations lead to the same conclusions, the movement of the waters of either is governed by the same laws.

When these observations on the Palmas were in course of progress we did not suspect that the percentage between surface and mean current would at the same locality materially change and be variable to a considerable extent; otherwise we should have had numerous mean current observations interpolated between the hourly surface current trials; the accepted rules which we carried with us admitted no room for suspecting the probability of such a change; and, accordingly, we concluded the observations by determining the percentage which the mean current bore to the surface current for a river like the Palmas, we thought at the time once for all.

Nor did the figures directly obtained from the trials call our attention to the variable nature of the mean current, expressed as a percentage of the surface velocity. We had, after the booking of the mean current trials, cast the figures up, to see approximately what the percentage on the Palmas amounted to, taking the "revolutions" per minute as a measure of percentage between surface and mean currents; not as absolute figures for velocities, but as relative figures for comparison. No diagrams were on hand to ascertain the surface current corresponding to the exact time of mean trial; taking, however, in both cases the observed surface currents, which promptly followed the mean trials, we found in the first the percentage to be 85?, in the second 83½ per cent., an agreement as near as could be expected, taking only "revolutions" per minute into account; and assuming the observed surface currents which followed the mean trials to represent the velocity during the operation of the current integrator.

The mean current trials on the La Plata and on the Palmas seemed at the time clearly to indicate, as far as the result of these observations could be considered without elaborate diagrams, that on the La Plata the mean current was about 75 per cent. of the surface current; and that on a river like the Paraná de las Palmas—which is but the Paraná on a smaller scale—the mean current would be about 85 per cent. of the observed surface velocity. There is no doubt that, at certain localities of the Plata, the Palmas, the Paraná, and other rivers, these figures would at times be quite correct; but to say that the mean current of a river or of an estuary would be so many per cent. of its surface current, is an imaginary view of the subject, and we have ample proof in the preceding pages that such expressions can never have any substantial foundation.

We now come to the second series of observations made at different points on the line of section across the Palmas, and which appear in Table V., under Nos. 45 to 51. All these observations were made within one hour and ten minutes, and it was a fortunate circumstance that the Palmas current was nearly constant during that time, otherwise, without the knowledge of the law of increase and decrease of the various currents, the curve representing the surface currents across the line of section would have either been distorted or the corrections of doubtful accuracy. The variations in the current of the Palmas having been but slight, a time was taken from which the increase and decrease of the preceding and succeeding observations were about equidistant, corresponding to 1.15 P.M., January 19th, 1871, and the corrections applied were but insignificant quantities. The exact position of the current observed on the river's surface was determined by triangulation, the mode of survey being fully described, pp. 20, 21. The trials from Nos. 45 to 51 give the surface current observed in feet per minute, and the distance of the specific current so observed from a point A, of the line of section on the left shore, being also a point of the base.

By plotting the calculated distances from A, or the complement of the angle observed at the locality of the current on A B from B, we obtain on the line of section a number of intersections fixing the position of each trial on the surface of the river. If, from the point of intersection so determined, we now draw a line in the direction of the current, and by the scale of the plan mark a point distant so many feet and inches from the intersection as the current was found to move in one minute, we obtain a point of the current curve across the line of section ; and by repeating this operation at all the points with their corresponding currents, and uniting them with a continuous line, we obtain a curve, marked on the diagram, "Velocity of currents in feet per minute across line of section as observed January 19th, 1.15 P.M., 1871."

The meaning of this line may be further illustrated by stating, that all the particles of water on the surface of the river, which at 1.15 P.M. happened to be at one time on the line of section, have, one minute later, occupied the various positions assigned to them by the curve represented on the plan of the survey of the Paraná de las Palmas, Plate IV. The curve representing the surface currents across the Palmas on line of section is most interesting ; we are, however, not yet prepared to enter into its significant outline because it is exceptional ; and, as we have not yet derived and shown the rule, it would be premature to consider the exception and trace the cause of disturbance.

Near the margin of the river the current is small, and increases as we leave the shore towards the middle of the river. The rate and mode of increase differs, however, materially, as we move from the right or from the left bank towards the middle. Within some localities the current remains stationary ; at others it increases and decreases rapidly ; at others, again, very slowly. It is a sinuous line of numerous bends ; and it is so with all rivers which flow within their natural irregular channels. The maximum or the minimum currents may be anywhere within the channel, and it will depend on the locality and the cross section of the river where they establish themselves.

The mean surface current across the Palmas at 1h. 15m. P.M., January 19th, on line of section from margin to margin of river, was 125 feet 8 inches per minute ; although it is a mistake to "mean" surface currents, for the result means not much ; it only extinguishes the peculiarities of the section and the locality. The reason is obvious. In the same river similar "mean surface currents" may be produced by essentially different sections, in the one of which we may have strong currents "meaned" with gentle ones, whilst at the other locality we may throughout have nearly an average current. The sections may be dissimilar in everything which distinguishes channels, inclination, and velocities, yet the mean may be similar; in short, it may mislead and can rarely be of much use. If it be necessary to compare surface currents, we may do so by naming the current corresponding to a certain depth and distance from shore; and with the knowledge of certain rules, the comparison will lead to absolute results, and the distinguishing features of the sections will not only be maintained, but more prominently shown.

## ANALYSIS OF SECTION.

The section of the Paraná de las Palmas at the locality of observations is shown on Plate IV. on a scale of 100 feet to the inch, horizontal and vertical dimensions. The drawing represents its natural form without exaggeration. How the soundings had been obtained and their exact position on the line of section determined will appear from the description of the mode of proceeding we followed in the survey of the Great Rivers, see page 11. Some of the sounding had been made on the 19th, in the evening; others next day in the morning; but with the gauge at the margin of the river they may be easily reduced to the same level, and in Table VI. the figures were given as obtained by the soundings, with the corresponding position of the Palmas level on the gauge. The distance of each sounding on the line of section from the base was determined by the sextant from the angle under which the base A B appeared, and calculated; and the result appears in a special column of the Table opposite each sounding.

The position of the Observatory O, about midway the river, is also shown on the section, the centre line of which happens to be within 8½ feet of the centre of movement of the superficial currents across the river; and, within 38 feet within the centre of gravity of the section. The width of the Palmas at low-water line—the lowest during observations—from margin to margin of water is 1165 feet; at high water the right bank is flooded, but the rushes determine the effective width at 1223 feet. The maximum depth of the channel at the lowest water (the datum of the diagram occurring at 3 A.M., January 19th) is 51 feet 9 inches, and for a considerable distance a depth of 50 feet is maintained. The mean depth of the Palmas at ordinary low-water level is 38 feet 9 inches for a width of 1165 feet from margin to margin of water, with a corresponding sectional area of 45,200 square feet. The high water above referred to, is not the level corresponding to the time of flood of the main river, the Paraná; it is the periodical high water due to the tides on the La Plata. At the time of the surveys, the Paraná was at its ordinary summer level, usually maintained for about seven months; the rivers were soon after commencing to rise, reaching their flood level about March.

The section shows the great uniformity of the channel of the Palmas, its symmetrical form, and even depth. It is as fine a section as could be desired for any artificial channel. The course of the Palmas at the section is slightly on the curve, the left bank being the concave side of its bed, on which the river also encroaches. About one mile above the section the bend of the Palmas is sharper in the same direction, the left bank remaining concave for some distance. The banks of the river are composed of stiff clay of deep-brown colour, and the soil appears to be very fertile. The level of the banks is not maintained for any considerable distance from the river; it falls lower and lower, and soon opens into a swamp, which frequently leads to an open sheet of water, a kind of inland lake, teeming with animal life. In the upper reaches, where the Palmas branches from the main river, it is rapidly shifting its bed, encroaching on one side and depositing on the other, maintaining however a nearly uniform width, depth, and sectional area; in the lower reaches its channel is almost straight, and seems to be settled and permanent.

# CHAPTER V.

## The Paraná.

### THE RIVER.

THIS mighty river is practically unknown in Europe. Although the Hydrographic Department of the English Admiralty possess ample knowledge of the river, as shown by their charts, the public at large, including engineers, know nothing about it. The Admiralty charts, the only reliable documents as a nautical survey in existence, are, however, little known, and rarely if ever used by those who navigate the Paraná. The navigation of this great river is invariably done by the aid of pilots, who do not look at a chart; and all the pilots we met with declared that they would not be guided by them, and some were candid enough to admit that they knew not what it was. During our expedition on the survey of the rivers, we always had one or two pilots on board the 'Aguila,' and we had ample opportunity to hear their views; to get at their ideas as to the rivers, and at their ways of piloting vessels. These pilots grow up in their profession, and navigate the Paraná from their boyhood; they know every tree and the outline and precise form of every "clump of trees" on the banks; and by these alone and the configuration of the shores are they guided in piloting a ship. They have a remarkable memory; and it is unquestionably a feat, to be fully appreciated by engineers, to see a pilot directing the movements of a fast steamer speeding along at from 12 to 14 miles as against the shore down stream in a dark night on a sea of water, sometimes not keeping five minutes the same course, and to ordinary eyes the shore invisible; the pilot, however, sees his way; shifting his course constantly in some of the upper reaches of the river, describing a number of sinuous lines in succession; and, the indirect proof that somehow he does see his way is given by the circumstance, that the moment a fog or a mist appears on the surface of the river, he stops the boat and drops anchor. If it be further considered that in the upper reaches of the river the channel is shifting within less than a year, and considerably so, it will be evident that in these localities charts could be of not much use. During our surveys, however, we found the Admiralty charts most valuable; all the features of a locality were readily understood and appreciated by the aid of the maps.

We now propose to take the reader on board the screw steamer 'Comercio de Paysandú,' and ask him to accompany us on one of the expeditions we made up the river. We shall enter the principal mouth of the Paraná, called "Guazú," and, being well defined by the termination of the left bank, we shall take it as a datum point for all the distances as we pass along up river; and as all distances will be measured from the same point, the difference between any two will be also the distance between the points. The mouth of the Guazú is in latitude 34° 0' 15" south, longitude 58° 24' 30" west, 10 English miles distant from a small island in the La Plata, called "Martin Garcia," which commands the entrance to the Paraná Guazú and to the Uruguay; it is a well-known point on the La Plata, and by adding 19 miles to all the distances on the river which we are about to give, they will reckon from Martín Garcia, the position and distance of which is known from all the towns and villages bordering the La Plata.

As we leave Martin Garcia for the Guazú, the horizon to the west appears a sea of water, and nothing is seen of the islands forming the delta of the Paraná. After about an hour's sail a fine line becomes perceptible,—a kind of fringe,—and we begin to see the margin of the La Plata. This is the visible delta of the Paraná, slightly above water, covered with luxuriant vegetation; the invisible delta, submerged, reaches close to Martin Garcia, and is daily growing; and in generations to come will surround this island of granitic formation. The Admiralty charts demonstrate it plainly, and even define to-day the form of the future outline of the delta near Martin Garcia. From these charts we know that the channel we are sailing in is about one mile wide, with steep sides, averaging between 30 and 40 feet depth; but the channel being submerged for many miles in all directions, it appears as if we were sailing on a sea. For the greater part of the distance from Martin Garcia the course of the steamer on the La Plata is N.N.W.; by degrees we approach closely the right bank of the Guazú; we then turn due west, and within half an hour from the time of change of course we are within the confined channel of the Guazú, which we enter on the 21st of May, 1872, at noon, having also come within the river's left bank; along its right we had already been sailing several miles. There are no banks visible even here; we see a thick growth of rushes and trees; there is, however, firm land on both sides, a couple of feet above the level of the river, but the rushes and branches of trees hide it; and unless we sail quite close to one bank, the land cannot be distinguished. The channel of the Guazú, as we enter it, is deep from bank to bank; the steamer may touch the boughs of the trees on either shore with several fathoms water below her keel. The width of the mouth of the Guazú is comparatively small; it is about half a mile; the Guazú having broken up into a number of separate branches, each forming a great river by itself, and all conveying jointly its volume to the La Plata and into the lake of the Uruguay. See Plate I., the Chart of the Rivers.

The scenery in the lower reaches of the delta is very fine, and there is a charm and grandeur in the profound stillness that seems to reign in these wild regions. The river glides along without a ripple or a whirlpool, and so smooth and even is the surface of the Paraná on a calm day, that we are more inclined to call it a lake than a river. The banks in the lower reach of the delta are covered with a thick forest of a peculiar tree called " Seibo "; this tree, which seems to have almost the monopoly of the lower reaches of the delta, is not unlike an oak deficient of leaves, having numerous stout branches of very crooked growth; and it would be difficult to find a straight piece of only 10 feet length either in the trunk or in the branches of a tree 60 feet high. It is rather short of foliage, the leaf being not unlike those of laurels; in the spring its flowers are as numerous as the leaves and of brilliant crimson colour, each flower of the size of a leaf; and the forest looks a mixture of dark green and bright crimson, certainly beautiful to behold.

As we speed on for miles and miles, there is no change perceptible; and in these tropical forests we have all along missed a population which claims the trees their home—

the birds; they are few and far between, and do not seem to inhabit them. The few we may now and then see are either carnivorous birds or waterfowl, and this may perhaps be accounted for by the circumstance that the delta produces little or nothing for other birds to feed on. We had at a time spent days and weeks along these forests, and had penetrated the woods some distance, not without difficulty; we always found the margin of the river deserted, but a short distance inland invariably marshy and sometimes an open sheet of shallow water covering square miles—we found the islands teeming with animal life of all kind; an abundance of wild ducks, geese, swans, turkeys, storks, cranes, snipe, &c.; and we call them wild, because they inhabit the wilderness; but, as a matter of fact, they are one and all very tame and may be closely approached, for, they know of no fear; they are never disturbed or hurt by anyone.

At 54 miles from the mouth of the Guazú we pass the line of section which we made in November, 1870, to gauge the volume of this branch of the Paraná. The section is a little below the junction of the Guazú and the Ibicuy, the latter returning the volume diverted by a branch called "Paranacito" from the main river at the head of the delta, see chart near 32° latitude; and conveys also the volume of another river called "Pavon," branching off a little above San Nicolas and joining the Paranacito. We had, moreover, gauged the volume of the Ibicuy independently, after the junction of the Paranacito and Pavon in January, 1871. The scenery remains about the same, except, that trees are getting less numerous, and that a long coarse grass covers the islands.

At 72½ miles from the mouth of the Guazú we pass the locality where the Palmas branches off from the main stream, which here divides into the Guazú and the Palmas Rivers. We now enter the Paraná proper; the bulk of the river being conveyed in one channel; its whole volume is only short of the quantity conveyed by the Ibicuy channel just mentioned.

At 98½ miles we sight the main land on the right bank of the Paraná, 70 feet above the level of the river. We are about two miles from the town of San Pedro, built on the main land at the margin of the bluff. The width of the Paraná is here 3950 feet, and the river is very deep with a strong current, and does not approach the town more than within one mile of its cathedral. The space between the bluff and the right bank of the Paraná is occupied by a lagoon—a kind of lake; it is on the average about three miles long and half a mile wide. This lake had been systematically surveyed by ourselves on Government account, and is the only natural harbour on the shores of the Paraná or the La Plata. Within the 18-foot contour line the available area of anchorage is equal to 312 acres, viz. within depths exceeding 18 feet at low water. This fine natural harbour is separated from the Paraná by a bank of clay, measuring only 295 feet width at its base from the 24-foot contour line of the lake to the same contour line of the Paraná; and yet this insignificant impediment had closed this convenient harbour to nearly all vessels navigating the Paraná, which up to this point may at the lowest level of the rivers draw 20 feet.

At 108 miles we reach Obligado, where the main land, about 60 feet above the level of the river, terminates in an almost vertical bluff, which forms the first time the right bank of the Paraná. The river here passes a kind of strait, its width narrowing to 2086 feet, or to less than one-half the average; the depth, however, increases correspondingly to 150 feet, the sectional area of river remaining about the same. At this locality a section was made with current observations in the month of December, 1870. The current about here is strong, and sailing boats are often delayed for days before they can pass the strait of Obligado. The high table-land of the province of Buenos Ayres now continues for more than a hundred miles to form the right bank of the Paraná, although by the intervention of large islands the deep channel of the present day is sometimes several miles distant from the main land. The left bank is an immense swampy island, about 100 miles long and 15 to 30 miles wide, commencing at the junction of the Guazú and Ibicuy, and terminating at the entrance of the Pavon channel. Hitherto we had been sailing among the islands of the delta, among luxuriant forests of Seibo trees; the main land being reached at Obligado, the scenery changes entirely; and although at the Strait the bluff is covered for a short distance with a wood, forming a fringe on the top of the barranca for a mile or so, trees become all of a sudden very scarce, and they are usually solitary Ombus; the table-land of the province being flat without any trees; a rich pasturage for tens of thousands of cattle.

At 148 miles we pass the town of San Nicolas, lat. 33° 20′ S., long. 60° 9′ W.; it is a rising place with considerable traffic, and, as we sail farther up river, the last town in the province of Buenos Ayres on the margin of the Paraná. At 152 miles the Pavon channel branches off from the left bank of the river to join the Paranacito, and thus to form the Ibicuy River. At 170 miles we enter a straight reach of the Paraná, about five miles long, of uniform depth and width; the right bank of the river being a barranca of from 70 to 80 feet high of the main land of the province of Santa Fé; the left is a sandy beach of the usual marshy islands of the delta. At this reach of the Paraná the great section of the main river was made by ourselves, January, 1871, with numerous observations; and, it is this section which disclosed and defined the law governing currents at the surface. The width of the Paraná at the section—distant 170 miles from the mouth of the Guazú— is from margin to margin of water at ordinary low level 4787 feet, and the depth increases from the left bank on a gentle and regular slope from a few inches to 72 feet 1 inch at a distance of 3720 feet from the left margin; or, at a distance of about 1100 feet from the vertical cliff of the right bank. At 3000 feet from the left bank, in a depth of 59 feet 1 inch, the surface current at the section was found to be 255½ feet per minute, or nearly three miles an hour. The whole volume of the Paraná at low water passes this section, excepting a comparatively small quantity conveyed by the Paranacito channel at the head of the delta of the river in lat. 32° 7′ S., long. 60° 35′ W.

At a distance of 183 miles we come to Rosario, the largest town on the banks of the Paraná, and of commercial importance. It is built on the margin of the bluff, about 80 feet above the river, and the town reaches from the main land down to the Paraná. Some desultory attempts at engineering stare at us in the shape of a number of strange structures, supposed to be piers; which a number of vessels anchored along the margin of the river seem rather to avoid than to court. Vessels drawing 15 feet may come up as far as Rosario at all times during the lowest level of the river. The rise of the Paraná from

o

ordinary low water to ordinary flood level is here about 12 feet, and the flood level is always maintained for at least three months. There are, however, periods when the flood is maintained during the whole year; and it does happen that the flood level remains permanently for two years in succession. The highest floods on record were about 24 feet above ordinary low water; during these the whole delta of the Paraná is submerged, and the river is one sheet of water from 15 to 30 miles wide. It then presents the ancient feature and outline of the former estuary of the La Plata. Such floods occurred in 1858 and 1868; in the former a vessel of considerable size came from the town of Victoria, on the Paranacito channel, to Rosario; a distance of about 30 miles in a straight line, the vessel sailing right across the islands of the delta with a cargo of lime, and drawing about 6 feet of water. The trade of Rosario appears considerable; we found invariably a number of large vessels at anchor, discharging or taking in goods in ancient Indian fashion. There is room for improvement and for engineers; and substantial progress seems to be checked only, and delayed indefinitely, by the unsettled local government of the province. Lat. 32° 55′ S., long. 60° 23′ W.

At a distance of 253 miles we reach Diamante, a little village, and the first time the left bank of the Paraná is bounded by the main land, a bluff of about 200 feet above the level of the river. The delta of the Paraná commences here in lat. 32° 7′ south. From Rosario the channel of the river is frequently changing and the navigation becomes more intricate and difficult. Below Rosario the channel does not change materially anywhere within a year or two; above the town, however, it is said that in certain reaches the channel always changes after the annual flood. A little above Rosario there is a large sand-bank in the middle of the river, and here the whole outline of the Paraná differs to-day from what it was at the time of the Admiralty surveys. For a long distance the main land of the province of Santa Fé continues to form the right bank of the river, and steamers sail closely to the vertical barranca, until, after several hours' sail, it retires and by degrees disappears from view; for a time we are again in the midst of islands. After a few hours' sail we sight the main land, but now it is on the left bank of the river; and we see the first time the Paraná bounded on the left by a bluff of considerable height, 200 feet above the level of the river called "Punta Gorda"; here the delta of the Paraná commences. For hundreds of miles the bluff of the main land now continues, with occasional breaks, to form the left bank; whilst the right is bounded by a succession of islands and the low-lying main land of the Chaco.

The bluffs, which plainly show the geological formation of the country, are of great interest; and, since we are compelled to sail for many hours within 100 yards of the remarkable formation which they exhibit, we may advantageously spend one hour with the geology of the country which this mighty river traverses.

The immense plain bounded between the Andes and the Atlantic, from west to east, and by the mountains of Brazil and the Strait of Magellan, from north to south, covering many hundred thousand square miles of land, is of tertiary formation, and the largest basin of the kind in existence. According to d'Orbigny this formation is composed of three distinct strata, of which the lowest he terms Guaranian; the middle one Patagonian;

and the upper Pampéan. The lowest stratum is a red sandstone, containing much iron and no fossils. On this rests a layer of a grey calcareous clay with small pieces of quartz and calcareous nodules. The upper layer is a grey gypseous clay, without iron. The middle stratum, or the "Patagonian," extends over larger areas, and the lowest layer is a grey sandstone of marine origin, containing mollusca of extinct species. On this rests another layer of sandstone, with a mixture of sand and clay with remains of mammalia and fossil wood. The next layer consists of calcareous sandstone, and agglomerates of marine shells, in which the Ostrea Patagonica predominates. The upper stratum, called "Pampéan," is partly a mixture of sand and clay, and partly pure clay, of red, yellow, and grey colour; which on the La Plata and Paraná is mixed up with masses of a hard calcareous clay, called "Tosca." The Pampéan stratum encloses numerous remains of mammalia, all of which belong to extinct races; it levels all the inequalities of the rocky surface below, and it is found in contact with the Silurian, Devonian, and other formations, as well as with the tertiary deposits above mentioned.

Our geological authorities seem to be agreed on the formation of the lower strata, which are of marine origin; but they differ as to the formation of the upper stratum, which is the land of the present day. It is a remarkable formation, containing numerous fossils and remains of the largest mammalia, which at a time were supreme over the land of the whole globe; and all the remains found belong to extinct races only. On the other hand, there is no trace of any marine deposit in this immense formation. We have carefully read the views and the hypotheses of d'Orbigny, Darwin, and others, and from the view which engineers would most likely take we incline to Darwin; and without expressing an opinion on geological matters, we will present the view we would take on the subject, guided by the opinion and the statements of these great savants.

The present plains of the Argentine Confederation, Bolivia, and of Patagonia, were at a time submerged and under a shallow sea; bounded by the Andes and the mountains of Brazil; and in this shallow sea the lower tertiary strata were forming. Comparatively slight geological disturbances raised and lowered in succession portions of this area above the level of the sea, and thus by degrees an inland sea, or estuary, of enormous dimensions was formed, substantially excluding the Atlantic. The sweet water which the torrents from the Andes and the mountains of Brazil poured into this shallow basin, was correspondingly displacing the sea water by numerous outlets, making the estuary brackish; and, in time, nearly sweet; which circumstance put an end to all marine formations within the estuary. The innumerable torrents from the mountain ranges of the Andes and those of Brazil were turbid and highly charged with argillaceous matter and fine sand, all of which the currents held in suspension as long as their velocity was maintained; as soon, however, as these torrents mingled with the still waters of the estuary, the matter held in suspension was by degrees deposited. The volume of water discharged into the estuary may at times have been also very different, not only depending on the rainfall, but also on the natural watershed of the mountain ranges, which at times may have drained into the "Old Plata Estuary"; at others, towards the present Amazon valley; so that not only different quantities of water were conveyed into the estuary, but also different quality, according to the nature of the lands which from time to time drained into it. Thus, at one period the torrents may have conveyed nearly pure quartzose sand and a comparatively small volume of water, which in a shallow sea, not yet confined, subject to tidal currents, would not materially alter the water of the sea, and admit the formation of marine deposits to a limited extent; at other periods the bulk of the volume of water draining from the mountains may have been altogether diverted from the shallow sea, and then we should expect an abundance of marine deposits; so that the variety of layers in the lower strata of this formation are readily explained by the frequent geological disturbances to which the region of mountains had been subject, and for which we have ample evidence by the numerous dislocations of the mountain ranges and of their masses. Darwin is of opinion that the tertiary deposits in the old Plata basin belong to one geological epoch, and not to several as suggested by d'Orbigny; originating, according to the latter savant, with violent disturbances in the basin itself principally caused by the formation of the Andes. D'Orbigny's views, apart from his eminent position, carry much weight; he having carefully studied the whole formation over large areas of the basin. Darwin closely examined some 300 miles of the territory, and especially "Punta Gorda," which we are just passing on our journey up river; he found in the lower strata a layer of red clay, with nodules of marl, identical with the uppermost deposit of the Pampas, and this layer was covered with calcareous matter, oyster-shells, and a variety of others of marine origin, and the whole again recovered by the usual Pampas deposit. This fact proves, in the opinion of Darwin, that the whole formation belongs to one geological epoch, and that the limits assigned to the layers of marine origin and to those of the clayey deposits are artificial. Darwin's discovery at Punta Gorda seems further to show, that the shallow sea bounded by the Andes on the west and by the Brazilian mountains on the north, was at times filled with salt water, at others with sweet water; and that the old Plata estuary was formed at a time and deformed at subsequent periods during the marine deposits; and again reformed and converted into an inland sea of brackish or even sweet water during the clay deposits and the formation of the present land of the Pampas. Nor does this involve improbable assumptions; considering, that during ages of time an insignificant variation in the level of these tertiary formations, extending over very large areas, might form an estuary by raising the surrounding lands near to the surface of the sea, or, by lowering the same lands, might again change the estuary into an open sea.

When the old Plata estuary was definitely formed, and in time its waters became brackish or even sweet by the torrents of turbid water which the Andes and the Brazilian mountain ranges poured into its basin, and marine life was extinct,—and over its former remains clay deposits were forming in the still waters of the estuary from matter held as long in suspension by the torrents as their currents were not materially checked,—we enter upon a new era of the same geological epoch; and we feel somewhat reassured that Darwin and d'Orbigny here agree, that the same torrents which filled the basin with their turbid clayey waters, also brought into the estuary the bodies of a race of mammalia which in that epoch flourished and reigned supreme in the tropical forests which covered the valleys and mountains of the land then above the sea level. We find that the clay deposits contain an abundance of remains of mammalia belonging to extinct races, and of a variety of species, such as canis, cerodon, milodon, mastodon, cervus, cebus, &c. &c., some fine specimens of which may be seen in the Museum of Buenos Ayres; and no other remains except those of extinct species are found. This is a significant circumstance, and determines the geological epoch of the Pampas formation as contemporaneous with the time when the great mammalia were in the height of their power and development; it was the age of the giant animals, of the mastodon,—the mammoth, the glyptodon, and others, then the reigning princes on the globe. The duration of their age, estimated by our authorities at about two millions of years, was abruptly terminated; their power was broken, and their race extinguished by another great change in the world's history; the Glacial epoch dawned, and changed the life and the organism of our planet. A chill of a hundred thousand years' duration had rather abruptly put an end to the great mammalia we have mentioned; and it is possible, if not probable, that their wholesale destruction—caused by atmospheric disturbances and by considerable changes of temperature producing exceptional torrents of melting snows from the high mountain ranges—had swept a great number into the old Plata estuary, converting it into an enormous graveyard of extinct races of mammalia.

During the Post-glacial epoch the whole of the old Plata estuary was raised by degrees above the level of the surrounding seas; and the various mountain torrents then united to form mighty rivers, which carved their beds in the deepest portions of the estuary; of which rivers the Paraná, Paraguay, Uruguay, are the great representatives of the present day. As these lands rose above the sea level, the vast extent of the old La Plata estuary was gradually reduced, until it was confined to the present delta of the Paraná and the La Plata of the day. All the animal life on these immense plains is but of very recent date, geologically speaking; and probably under two hundred thousand years of existence. It may hardly be necessary to observe, that there must have been many changes, and that the rise in some places preceded others, and so on; and it should not be supposed that these events occurred in the simple outline we have indicated; indeed, these events, spreading over millions of years, were in all probability very numerous and complicated in their consequent reaction, and all we can do is to summarize and note a few main events which ultimately may have determined the outline of the great South American continent as it now exists. We shall, in conclusion, allude to a remarkable formation of an argillo-calcareous rock called "Tosca," which interlaces in all directions the thick deposit of clay of the Pampas. As a rule it is only met with in most irregular pieces varying from a few ounces to many pounds weight each. It is found as well at the surface as anywhere within the clay deposits. A hardened kind of reddish clay is, in popular language, also called tosca; but the real tosca is of grey colour, and appears like veins within the hardened mass of the clay; and good specimens of this kind of formation may be seen on the beach at Buenos Ayres. Its chemical constitution varies considerably; one half of the substance may be taken as carbonates of lime, magnesia, and iron; and the other half as silica and alumina; one half of the silica being combined, the other in excess, and in the shape of a fine sand. It may be termed a silicious compact marl. All this, however, throws but little light on the origin of these substances. At San Pedro the formation may be seen to advantage, and here the bluffs exhibit the mode of "growth" of tosca. The barranca at San Pedro is 70 feet high, and as the rain by degrees removes and washes away the soft clay deposits not yet hardened in this district, we see beautifully the skeleton of the tosca formation. At the foot of the bluff at the eastern end of the Laguna, there is a stratum of perhaps 2 feet thickness, within which we have dense masses of tosca interlacing each other in all directions, of most irregular form, with numerous cavities equally irregular; all, however, grown together and sometimes uniting to a solid layer of a couple of inches in thickness. From this comparatively dense stratum of tosca shoot, in a vertical direction, numerous branches, from half an inch to one inch in thickness, round; similar in appearance to the roots of plants and trees, for which we mistook them from a distance at the first sight. The branches of tosca, retaining the same thickness and form, and running nearly in a vertical line from 3 to 6 feet and more, again join another dense layer of tosca, perhaps only 1 foot in thickness, similar in every respect to the lower one already described. The spaces between the branches and the cavities of its denser layers are throughout filled up with clay. Looking at this formation in its original condition, tosca cannot be called a "deposit" in the ordinary sense of the word, no more than a tree which buried at the locality where it had grown could be called a deposit. The formation of toscas appears similar to that of corals, and the toscas seem to be the produce of stone-making polypes. All coral animals require, however, salt water; and it seems to be even of greater importance that the water in which they live and build should not be turbid either from surf or from matter held in suspension; the latter being most injurious to the existence of all zoophytes. The appearance of tosca, when its original outline is not obliterated by wear and tear, greatly resembles the stony branches of the species of Porites and Madreporæ; though, it seems, that at different periods different species of zoophytes inhabited the Plata basin whilst as a shallow ocean it was free from the sediment of rivers; because the dense strata of tosca closely interlacing and sometimes uniting to solid masses, are quite different in appearance from the long thin branches which follow it in other layers. The former are porous, the latter solid, all of which are of a grey and light-brown colour. We broke from the dense layer a lump of tosca in which a cell was entirely preserved, and others partially. The opening at the top of a conical projection was perhaps ½ to 1/16 inch diameter; circular with a sharp margin, the cavity rapidly increasing in diameter to perhaps ½ inch at a depth of about ¼ inch. This cavity was in all probability the cell in which the polype lived. We had carefully examined the little projecting cone, and found it to be the same substance as the lump of tosca, which weighed about 10 lbs. At one period the porites and other species of coral polypes inhabited the shallow ocean of the Plata basin, then extending from the present mouth of the Plata estuary to the Andes. It was after the deposit of the stratum called by d'Orbigny "Patagonian"—which he describes as a grey sandstone of marine origin, containing mollusca of extinct species, and in the upper layers calcareous sandstone and agglomerates of marine shells—that the reef-building corals appear to have monopolized this shallow basin filled with sea water, and produced a submerged forest of stony reefs; when, by the gentle rise of the adjoining bottom of the sea, a vast estuary was formed, from which by degrees the salt water was displaced by the torrents of the mountains of Brazil and the Andes; which, with their turbid water, have not only terminated the existence of the zoophytes, but by their deposit have also buried the skeleton of the reefs, and even filled the finest pores in which once the zoophytes had lived with impalpable sand and clayey matter. What then in popular language is called Tosca, appears to be the remains of an ancient coral-reef.

After this geological digression we sight the town of Paraná at 288 miles from the mouth of the Guazú, in lat. 31° 44′ S, long. 60° 25′ W., on the top of a bluff on the left bank of the river. It is the capital of the province of Entre-Rios, and the most ancient settlement on the Paraná. About twenty years ago it was temporarily made the capital of the Confederation, the whole country being at the time greatly disturbed. Vessels drawing 12 feet may at all times come up as far as the town of Paraná. From here the bluff, about 120 feet high, frequently forms for many miles the left bank of the river, and from

Feliciana to La Paz, a distance of 30 miles, it is the bank of the Paraná without intermission.

At 357 miles, in lat. 30° 58½' S., long. 59° 42' W., we come to the first pass, called "Feliciana," and about here the river begins to assume a wild appearance; with numerous islands and sandbanks.   At 368½ submerged rocks are crossing obliquely the bed of the river, which loses its calm appearance, its surface being agitated by a number of whirlpools; and the immense mass of water over a square mile of area appears in great commotion, as if in convulsions.   There is a strong current about here and difficult to pass for sailing vessels, although there is no danger to navigation from these rocks, which are several fathoms below low-water level.   Five miles higher up, at 372, a similar ledge of rocks crosses the channel of the Paraná, over which there is deep water with a strong current.

At 389 miles we reach the town of La Paz, on the left bank; all the towns and villages higher up the river will be on the left, and none on the right; the latter consists either of a succession of islands or the main land of the "Chaco"; a vast tract covered with a dense forest, entirely in the hands of savage Indians.   There may be here and there isolated spots reclaimed by civilization in the shape of "Colonies," but they are few and far between. From La Paz, in lat. 30° 44' S., long. 59° 30' W., the river is getting more and more wild and irregular, with many shallows and numerous large islands covering many square miles each.   Close to the town the lower pass of San Juan commences,—then follow the middle and the upper passes at 405½ miles distant from the mouth of the Guazú.   Above the San Juan Pass the river is again assuming a more regular shape.   At 447 miles we come to Esquina, in lat. 30° 2' S., long. 59° 24' W.; the town being about four miles from the Paraná, the channel of which was nearly due north since La Paz, and the main land far away from both banks the whole distance; there are occasional shallows of 10 feet at low water.   At this station, on the right bank of the river, one of the largest colonies had recently been established by a first-class London financial house.   We had occasion to visit the Alexandra Colony, and it was an agreeable sight and relief to find in the midst of the wilderness, among hostile Indians and tigers, and worse than tigers,—among myriads of mosquitos and insects of all kinds the chief plague of the land—thrashing machines, steam flour-mills, and agricultural implements, with about 500 hardy Europeans as the pioneers of civilization; opening new territories for the expansion of their race and the future generations of our over-crowded Old World.   From 484 miles to the town of Goya, the Paraná is getting very wild, and this is the worst reach of the river at present under navigation.   The Paraná here breaks up into a number of great islands, covering from 5 to 30 square miles each; with numerous smaller ones and still more numerous banks, which change position.   This reach includes the Jaguaraté Pass, the worst on the river, at 508 miles from the mouth of the Guazú.   The main land is distant from both banks, which are a succession of islands.

At 530 miles we pass the channel leading to the little town of Goya, in lat. 29° 7' S., long. 59° 13' W., the Paraná continuing its wild unsettled course.   In this reach of the river we had, at 3 A.M., May 25th, 1872, to drop anchor on account of a dense fog, which at 6 A.M. melted away before the power of the rising sun, having been delayed three hours; the engines were stopped the first time since we left Monte Video on the 20th May, at 4 P.M.,

entering the channel of the Guazú on the 21st at noon, on our up-journey of the Paraná.
At 541 miles the main land forms the left bank for about 12 miles, and again from 567 to
584 miles; at which point we come to the town of Bella Vista, in lat. 28° 29' S., long.
58° 58' W. Here the navigation again becomes difficult, the river breaking up into a
number of great islands. The course of the Paraná is for a considerable distance on the
59th meridian. At 593 miles there is a pass called "Tres Bocas," with only 7 feet at low
water, and at 635 miles the main land again forms the left bank as far as Corrientes, distant
666 miles from the mouth of the Guazú. The town of Corrientes, in lat. 27° 27½' S., long.
58° 44' W., in the province of the same name, is of some importance, situated a little below
the confluence of the Paraguay with the Paraná. At Corrientes the Paraná is confined to
one channel of over a mile in width, of great depth and a strong current. The river looks
here rather more powerful and greater than below Rosario, and its volume, if anything, will
be greater here; it may lose more by evaporation during its long journey to the La Plata
than gain by the insignificant tributaries in the hot summer of this region, so deficient of
rain and moisture. The position of Corrientes, at the confluence of the Paraguay with the
Paraná, is favourable; and vessels drawing 7 to 8 feet may at all times during the lowest
condition of the river come up as far as Corrientes, and during the period of flood they may
draw 10 to 12 feet more. Corrientes commands the highway of two great rivers, which,
coming from different regions of the Central American continent, here join. For the
present, navigation higher up is confined to the Paraguay; a fine river, navigable for 1000
miles, whilst the navigation of the Paraná is not attempted, for reasons we shall presently give. The situation of Corrientes is also favourable for an arsenal and a dockyard,
commanding the trade of the Paraguay and the upper Paraná; and, on account of the
excellent timber with which the Grand Chaco on the opposite shore abounds, enough
material might be found for the construction of any number of fleets for the expanding trade
on these magnificent rivers. The channel of the Paraná, which for the last 300 miles had
been N.N.E. and north, here changes rather abruptly to east, in which direction the
river continues its course for a couple of hundred miles; we do not, however, continue
our journey higher up the Paraná, for our steamer 'Comercio de Paysandú' is bound for
Assuncion, the capital of Paraguay; and within three hours' sail from Corrientes we touch
the island of "Cerrito," lat. 27° 19' S., long. 58° 32' W., at 683 miles; on the right bank at
the mouth of the River Paraguay; and we land on the 26th May, 1872, at 11 A.M., the
first time since we left Monte Video, 20th May, 4 P.M., after a net run of 136 hours up river
for a distance of 852 English miles.

We asked the reader to accompany us on our journey up the river as we left Martin
Garcia on the La Plata, and we took him on board the 'Comercio' without, however,
telling him that we were not only declared by the Argentine Government to be in
quarantine, but that we were entirely excluded from their territory; even if we had been
willing to submit to any number of days of quarantine, because we had sailed from
Monte Video, where there had been daily perhaps half-a-dozen cases of yellow fever.
The anxiety of the Government to take stringent measures against its importation into their
unhealthy capital may be readily appreciated, especially since Buenos Ayres is deficient of
ordinary sanitary arrangements and regulations, and without sanitary works of any kind;
their anxiety is the more natural since the city had been visited but twelve months previous
to the appearance of yellow fever at Monte Video, with the most terrible Yellow Fever

epidemic on record in modern times; we believe there was ample ground for serious alarm; the city having been in the same condition. We understand that contracts for the execution of sanitary works for Buenos Ayres have recently been concluded. We had ourselves been in the midst of the epidemic at Buenos Ayres in February, March, and April, 1871, decimating the city in a couple of months; thousands of victims had rapidly filled the old cemeteries; a line of railway had to be made in all haste for the conveyance of the thousands who, at one period, disappeared every week as if by magic. By excluding and cutting off all communication with the infected sister-town the authorities erred on the right side; but we do not believe that the extreme measures which had been adopted in consequence of the serious circumstances had saved their city from another epidemic. That it did not make its appearance at Buenos Ayres in 1872, may, to some extent, be due to atmospheric changes; but in all probability it was chiefly due to the circumstance that people inclined to fevers had been carried off in 1871. A fire also ceases to burn for the want of fuel; and a conflagration may end without suppression by artificial means. We now disembark the reader at the "Cerrito," at present in the hands of Brazilian forces, and a kind of military station and arsenal; the land is claimed and belongs to the Argentine Confederation; having, however, touched a Brazilian station, we are permitted to take any vessel or steamer and cross over to Corrientes, or any other part of the Republic, and enter any town without even the attending nuisance of a quarantine,—all of which is a sharp turn in the opposite direction of the extreme measure to which we had been subjected only six days ago, excluding us entirely from Argentine territory. The exploration of the Paraná is, however, by no means completed; for, the river at Corrientes is quite as large as at Rosario; its navigation higher up is for the present abandoned, and we cannot conveniently further explore it, and must content ourselves with statements and accounts of travellers of ancient and modern times.

Above Corrientes the channel of the Paraná is on both sides confined by main land considerably above its level; the banks rising higher and higher as the river is ascended, and the region of mountains is more and more approached. The channel is no more divided by numerous large islands as below Corrientes; they are few and far between, small and rocky, and the character of the ground which the river traverses is changing from alluvial deposit to the firm outline defined by rocks. The aspect of the river is grander in this reach than below Corrientes, because here the Paraná may be seen as a whole, which is rarely the case in the lower reaches, where we can but see a part of it between a succession of islands. At about 150 miles above Corrientes two large islands, called Apipé and Yaureta, separated from one another by a narrow channel, are within the Paraná, and here a ledge of rocks crosses its bed, forming rapids; one ledge reaching from the island of Apipé to the main land of Corrientes, and the other, produced by the same bar of rocks, is between the island of Yaureta and the main land of Paraguay; the former is called the Fall of Apipé; the latter, and the higher of the two, the Fall of Aregua. During the low-water season of the Paraná these Falls cannot be passed by vessels, but during flood time the bar of Apipé is submerged, forming a short rapid, which may be passed by steamers of small draught. After these rapids the ruins of Itapua on the right bank and those higher up of Candelaria on the left bank are reached; and it is said the scenery near Itapua, a hilly district covered with dense tropical forests, is magnificent. At the ruins of Candelaria, 186 miles distant from Corrientes, the river narrows to less than one-half its ordinary width, which it

P

soon regains, at the same time changing its course, which hitherto had been east, to N.N.E. ; the Paraná now entering the region of the Brazilian mountain ranges. The territory of the ancient " Missions " here borders the Paraná, of which the ruins of San Ignacio, Loreto, and others, are the only evidence that once the hand of civilization had reached and penetrated into these distant parts of the world, where now-a-days not a vestige of civilization remains. The wild and luxuriant tropical forests are in the hands of a few wild and naked Indians, who, terrified at the sight of a white man, fly at his approach. In ancient times the Jesuits had done also good ; they had penetrated into the heart of the South American continent, and had spread the light which reclaimed the Old World from the hands of heathen barbarians. In 1631 their settlements of Guaira in the upper reaches were destroyed by the Mamelukes of St. Paul of Brazil, and Father Montoya retired lower down and formed the new settlements of Corpus, Loreto, Santa-Ana, and others. The Jesuits had excluded the Spaniards from the new settlements, the population of which consisted entirely of Guarani Indians. The new settlements below the Great Fall of the Paraná, called "Guaira," reached an unprecedented prosperity about 1750, which aroused the jealousy of the other Spanish provinces of Buenos Ayres, Paraguay, &c. The population of the Missions then comprised about one hundred thousand Guaranis under the absolute government of a comparatively small number of Jesuits of French, German, and English extraction, all good men and well educated ; and they achieved a great success. They were nominally under the authority of the Spanish Crown. The Portuguese of Brazil coveted the Missions ; and, Spain surrendered them to Portugal by a treaty in 1750 ; in consideration of an insignificant Portuguese settlement on the La Plata, called " Colonia," which affected injuriously some Spanish merchants by the contraband of the Portuguese. The Jesuits were expelled from the Missions soon after the fall of their Order in Europe, although they were respected and beloved by the Guaranis, who then came under the rule of Spanish governors and Franciscan monks ; from which time the Missions rapidly declined. The population was in thirty years reduced to less than one-half, and they were continually involved in Spanish troubles and wars. In 1817 the Portuguese Governor of the province of Rio Grande sent one of his generals to surprise the Guaranis and to destroy everything in the Missions by fire and sword, and to carry the remnant of the Indian population to Brazilian settlements. The orders had been faithfully executed, and the Missions were barbarously destroyed. A century ago they were the most flourishing settlements on the continent of South America. The Guaranis were a docile race of Indians of mediocre intelligence. Their settlements under the Jesuits were most remarkable—there was absolute equality and no classes ; all had to work and all enjoyed equally the fruit of their labour ; and they formed a community not unlike the bees of a hive. To-day the territory of the Missions is a wilderness, out of the pale of civilization.

According to accounts of missionaries and Castelnau, the first great tributary falls into the Paraná above the Paraguay, in lat. 25° 45′ S., called " Y-Guazú," a great river of the first order ; a little higher up a number of secondary rivers on either bank join the Paraná, pouring immense volumes of water into its channel, which in this reach is broken up by numerous ledges of rocks forming a succession of rapids, all of which may be passed, not without difficulty, until the great Fall of the Paraná called "Guaira," in lat. 24° 4′ 38″, puts a final stop to all navigation. This is a remarkable Fall, and according to recent and ancient descriptions its grandeur can hardly be conceived. The first account we had of the Fall of Guaira was from two gentlemen, a wealthy Estanciero of Corrientes, and from a General of cavalry of the army of Lopez; and the source deserved confidence. We had nevertheless doubts about its general accuracy, as it is not unlike favourite stories of the type of the 'Arabian Nights,' and we should have omitted it altogether had we not found the account of the recent explorer generally confirmed by the description of men of education, who officially visited these regions eighty-five years ago. The statement of the Estanciero and of the General was as follows:—In the year of 1863, Don Carlos Lopez, Dictator of Paraguay, sent Domingo Platiño, an officer of his army, and about fifteen soldiers, to explore the right bank of the Paraná as far as the river borders his dominions; and, with great difficulty and much privation the exploring party reached as far as the great Fall of Guaira, of which Platiño gave the following description to Lopez :—

"The Fall of Guaira is about 230 leagues (say 700 miles) from Corrientes up the Paraná, and the thunder-like noise of the Fall could be heard at a distance of 10 leagues (30 miles), and within one league the noise was so great that it was difficult to hear one another; at the Fall itself it was impossible to distinguish any voice, all being drowned in the thunder of the terrific concussion of contending waves. It was not a vertical drop which caused the thundering noise, but the concession of immense waves which broke upon vertical granite walls, narrowing the ordinary width of the Paraná to about 70 mètres; the difference between the upper and lower level may be 20 mètres. It is reported that settlements close to the Fall of Guaira had to be abandoned on account of the noise, which made the population deaf. The whole of this region is at present in the hands of the wildest class of savages, a miserable race of Indians, of which they had caught one as a sample."

Making due allowance for the estimate of an illiterate man as to distances and dimensions, there is no doubt that the officer faithfully reported what he had seen, well knowing that any error or untruthfulness would, at the hands of his tyrant chief, have been at the expense of his life. The latitude of the Guaira Fall having been determined by scientific explorers, its distance from Corrientes is about 150 leagues, or about 450 miles; and not 700, as estimated by Platiño; and we may probably halve his other distances where he speaks of leagues; all of which will, however, not materially affect his account. The Commissioners of Limits in 1788 visited this prodigious Fall, which goes also by the name of Maracayu, on account of a chain of mountains of the same name which cause it by crossing the channel of the Paraná. The Commissioner Azara gives us the following written description :—

"Imagine an immense Fall of water worthy of the description of poets; for it refers to the river Paraná, a river which, even in this locality, 470 leagues from its embouchure, has more water than all the great rivers of Europe united, and which near the Fall still has a width of 4200 mètres. This immense width is all at once reduced to a narrow channel of 60 mètres, in which its waters break with indescribable fury. Its waters do not drop in a vertical line, but upon a plane inclined about 50 degrees, forming a clear drop of 17 mètres measured in a vertical line ; the clouds produced by the concussion of the water against the walls of its granite channel and on the rocks which project in the middle of the current form columns of steam which may be seen for several leagues, and on which numerous rainbows are visible. A continuous rain, produced by condensation, falls in the neighbourhood ; the noise of the cascade is heard six leagues distant, and close to it one believes the earth is trembling. 1788. Azara."

There is nothing contradictory in the two accounts given by two very different persons at different times, seventy-five years apart. One statement substantially confirms the other, although, in detail, each explorer treats the subject in his own way, as two persons would who knew nothing of each other. That Platiño gives his distances in mètres is not strange if we consider that his chief, Lopez, was educated in France, and in his land the Dictator practised French habits and customs. We would, nevertheless, pause before we accept the

P 2

figures of either explorer; they are probably estimates which, in the intense excitement the prodigious scene cannot fail to produce, are naturally overdrawn. Nothing short of a survey can, under similar circumstances, determine distances, outline, and masses; all of which must be matter for observation and not for estimate. What may be taken for granted is this,—that a prodigious Fall, involving the concussion of immense masses of water, exists in the Central South American continent on the Paraná, the largest river in the world, except, perhaps, the Amazon; that the Fall of the Paraná is over a steep, rocky incline, and that its channel is abruptly and very much reduced and confined within walls of granite; that the concussion must be tremendous; for, unlike the Niagara—where the whole mass, in tumbling over a precipice, is received in the deep water of a large basin, in which it is buried and disappears from sight—at the Guaira Fall of the Paraná the whole work of the falling masses seems to be expended upon rocks, there being no basin to receive and to soften the concussion of a million tons of water per minute; racing along at a speed of probably above 40 miles an hour, and immediately broken up by a succession of terraces and walls of granite.

Above the Fall of Guaira, as far as the Rio-Pardo, another great tributary about 270 miles distant from the Falls, there is little or nothing known of the Paraná; except that, according to ancient missionaries, there are a great many rapids within its channel. As far as latitude 22° 30′ S. the course of the Paraná since Candellaria had been north to south; here it turns to N.E., receiving all along a great many large tributaries, and, according to Castelnau, its source is in latitude 16° 30′, near the town of Goyaz, within a range of mountains of small elevation, where the little stream is called Corumba, assuming, near the village of Bomfin, already considerable dimensions from innumerable torrents which shed their water into the Corumba; it then receives the Paranahiba, not inferior to the Corumba, and a number of others; and, lastly, the Rio Grande, from the junction of which it takes the name "Paraná," which, from its sources to its embouchure in the Plata estuary, extends over a distance of about 1800 miles. It also appears, that the volume of the Paraná is principally, if not wholly, derived from the mountains of Brazil, and that in the present epoch the Cordilleras supply little or nothing to swell it. On the maps numerous torrents seem to descend from the Andes into the great plains of the tertiary basin; but most of these seem to lose themselves within swamps; at any rate, the Paraná receives but insignificant supplies from the far west; for, even the Paraguay, which joins it at Corrientes and to its west, principally drains the mountains of Brazil, and not the Cordilleras of the west coast, which separate the plains from Chili. We may except the Vermejo and the Pilcomayo, which, drain Bolivia in the west; but their volume is at all times insignificant in comparison to that of the Great River. The bulk of the volume of the Paraná is derived from Brazilian mountain ranges within the tropics; from a great number of large rivers and innumerable torrents which unite to make up the mighty river on the South American continent. Its drainage area may be estimated at so many hundred thousand square miles, but it is idle to define it by figures until its tributaries shall be a little known, and the rainfall of the respective regions shall have been measured. At Corrientes the Paraná drains about half-a-million square miles of mountainous regions, and it is these which supply the bulk, if not substantially its whole volume; for, it is doubtful whether the insignificant tributaries during the summer months will compensate the evaporation attending the exposure of over

**2000 square** miles of the Paraná's surface from Corrientes to the La Plata. Geographically **speaking,** nearly the whole of the La Plata basin, bounded by the Cordilleras on the west, **Bolivia and** Paraguay on the north, and Uruguay on the east, must also drain into the Paraná—an area larger than its mountainous watershed ; yet this immense area supplies next to nothing ; first, because the rainfall is comparatively small, and within whole regions nothing ; second, because this part of the watershed is almost level, and before it could drain into the Paraná it is absorbed by the ground and subsequently taken up by vegetation **and** partly evaporated ; third, because heavy rainfalls on the Pampas collect in swamps **and** lagunas, or inland lakes, from whence most of the water is again evaporated during **the dry** season. The drainage area of a river like the Paraná, which **traverses** regions **of primary,** secondary, and tertiary formations ; **mountain** ranges, and **alluvial plains,—should not be** expressed by **a** geographical measurement **of the area which ought to drain into it according** to inclination, **because such figures would be greatly misleading and useless for comparison ;** watersheds, which are entirely different in character and formation, should be separated, and their rainfall and supply treated independently ; and then we may form an idea from whence the river derives its power and volume, and which are the regions which principally affect its low and its flood water levels.

## THE ROSARIO SECTION.

In the preceding pages we had occasion repeatedly to refer to the great section of the Paraná, not far from the town of Rosario, in the province of Santa Fé. It is probably the largest section on record ; it is within a straight reach of the river, and of an outline rarely to be met with anywhere. It is this section which disclosed, defined, and determined the law which governs the velocity of surface currents. This law may, after its discovery, be easily recognized in the other great sections of the Paraná and Uruguay, but it could not have been determined from any of the others ; because their geometrical form makes the law vague, which, confined within a small range of variation, makes its operation at greater distances uncertain. We have already, in the second chapter, given a detailed description of the locality of the Rosario section, and the mode of proceeding we had adopted to determine the soundings ; the velocity of surface currents from bank to bank on line of section ; the relative position of every observation on the surface of the Paraná, &c., for all of which we now refer the reader to pages 10, 11, 12, and 19, 20, and we shall at once proceed to give the entries of the Survey Book referring to observations of the Paraná, near Rosario, at the locality of section.

### DIARY OF OBSERVATIONS.

JANUARY 23RD, 1871. LOCALITY: PARANÁ ; LEFT BANK, LAT. 33° 5' S., LONG. 60° 25¼' W.

Section about 12 miles below Rosario. Base, 3000 feet on left bank and near margin of river ; magnetic bearing of base looking up river, 303° 30'. Base at point A of section, 83 feet 5 inches from margin of river. Sandy beach gently sloping into river. Rushes end at 142 feet from A in water 2 feet deep. Angle from C on right bank on line of section at the vertical bluff, 31° 38' on base A B. Line of section at right angles with base. Direction of right bank looking up river, 311° by prismatic compass. From point C at bluff on right bank, to margin of river, 59 feet 2 inches on line of section. A flood level visible on the bluff from 21 to 23 feet above present level. Bluff 59 to 60 feet above river. Gauge on left bank 2 feet 4 inches. Noon, January 23rd, 1871.

JANUARY 24TH, 1871. LOCALITY: PARANÁ, ROSARIO SECTION, SOUNDINGS.

*Soundings on Line of Section*, 11 A.M. (from board the Steamer ' Aguila ').

| No. of Observation. | Angle at Locality of Observation on Base A B. | Sounding in Feet and Inches. | Remarks. |
|---|---|---|---|
| | ° ′ | ft. in. | |
| No. 1 | 81 34 | 16 7 | Commencement, 11 A.M. Gauge, 2 feet 5 inches. |
| „ 2 | 76 23 | 27 5 | |
| „ 3 | 71 43 | 37 9 | |
| „ 4 | 66 5 | 42 0 | |
| „ 5 | 60 12 | 49 0 | |
| „ 6 | 54 9 | 53 8 | |
| „ 7 | 50 8 | 58 0 | |
| „ 8 | 48 36 | 58 1 | |
| „ 9 | 45 0 | 59 1 | |
| „ 10 | 42 8 | 69 6½ | |
| „ 11 | 40 27 | 68 9 | |
| „ 12 | 37 49 | 24 5 | {After this sounding changed course on account of great difference in depths. |
| „ 13 | 40 55 | 68 5 | |
| „ 14 | 40 12 | 70 11 | |
| „ 15 | 38 38 | 72 1 | |
| „ 16 | 37 8 | 12 4 | |
| „ 17 | 35 41 | 9 1 | Soundings terminated, 1.16 P.M. |
| C. | 31 38 | —6 0 | |

NOTE.—Fine weather, gentle breeze, heat intense, oppressive.

JANUARY 25TH, 1871. LOCALITY: PARANÁ, ROSARIO SECTION.

*Surface Currents on Line of Section across River.*

Meter No. 1. Perfect calm.

**No. a.** Position of observation : Angle on A B, 80° 33′.

Time {Commencement, 7h. 50m. 0s. / Termination, 7h. 55m. 0s.} 5m. 0s. observation. Indexes {(0a + 0) / (2a + 10)} 412 revolutions.

Check observation     1m. 0s.     „     „ {(2a + 76) / (2a + 156)} 80 „

**No. b.** Position : floating flag, Angle on A B, 74° 31′.

Time {Commencement, 8h. 18m. 0s. / Termination, 8h. 23m. 0s.} 5m. 0s. observation. Indexes {(0a + 0) / (3a + 7)} 610 revolutions.

Check observation     „     „     1m. 0s.     „     „ {(3a + 7) / (3a + 128)} 121 „

**No. c.** Position : Angle on A B, 67° 37′.

Time {Commencement, 10h. 8m. 0s. / Termination, 10h. 13m. 0s.} 5m. 0s. observation. Indexes {(0a + 0) / (4a + 20)} 824 revolutions.

Check observation     1m. 0s.     „     „ {(4a + 20) / (4a + 180)} 160 „

**No. d.** Position : Angle on A B, 66° 7′.

Time {Commencement, 11h. 15m. 0s. / Termination, 11h. 20m. 0s.} 5m. 0s. observation. Indexes {(0a + 0) / (4a + 133)} 937 revolutions.

Check observation     „     „     1m. 0s.     „     „ {(4a + 133) / (5a + 115)} 183 „

**No. e.** Position: Angle on A B, 60° 34'.

Time {Commencement, 11h. 48m. 0s.} / {Termination, 11h. 50m. 0s.}   5m. 0s. observation.   Indexes {(0a + 0)} / {(5a + 32)}   1037 revolutions.

Check observation .. .. .. 1m. 0s.   ,,   ,,   {(5a + 32)} / {(6a + 40)}   209   ,,

Note.—The boat anchored fore and aft oscillates a little, between angles 60° 34' and 60° 40' on line of section.

---

**No. f.** Position of observation: Angle on A B, 54° 2'.

Time {Commencement, 12h. 21m. 0s.} / {Termination, 12h. 26m. 0s.}   5m. 0s. observation.   Indexes {(0a + 0)} / {(5a + 161)}   1166 revolutions.

Check observation .. 1m. 0s.   ,,   ,,   {(5a + 161)} / {(6a + 196)}   236   ,,

---

**No. g.** Position: **Angle on A B, 46° 48'.**

Time {Commencement, 1h. 5m. 0s.} / {Termination, 1h. 10m. 0s.}   5m. 0s. observation.   Indexes {(0a + 0)} / {(5a + 82)}   1238 revolutions.

**Check observation** .. .. 1m. 0s.   ,,   ,,   {(5a + 32)} / {(7a + 78)}   247   ,,

---

**No. h.** Position: Angle on A B, 40° 17'.

Time {Commencement, 1h. 36m. 0s.} / {Termination, 1h. 40m. 55s.}   4m. 55s. observation.   Indexes {(0a + 0)} / {(6a + 0)}   1206 revolutions.

**Check** observation 1m. 0s.   ,,   ,,   {(6a + 0)} / {(7a + 43)}   244   ,,

---

**No. k.** Position: Angle on A B, 37° 42'.

Time {Commencement, 2h. 10m. 0s.} / {Termination, 2h. 15m. 0s.}   5m. 0s. observation.   Indexes {(0a + 0)} / {(2a + 104)}   506 revolutions.

Check observation .. .. .. .. 1m. 0s.   ,,   ,,   {(2a + 104)} / {(3a + 8)}   100   ,,

Note.—Boat anchored fore and aft oscillates between 37° 41' and 37° 42'.

---

### MEMORANDUM.

Fine bright day ; a perfect calm. Surface of Paraná like a pond, without a ripple. In the greater depths some difficulty to moor the little boat fore and aft, the fore anchor dragging in the strong current. The two small boats used during current observations towed into position by 'Aguila.' Difficult to estimate distances for dropping anchors. The heat intense, 110° Fahr. in the shade ; a hot day's work. Temperature of Paraná water, 80° Fahr. At 3 P.M. sailed for Rosario to meet the pilots at the Captain of the Port.

## ANALYSIS OF OBSERVATIONS.

The preceding observations may now be worked out in **a similar manner to those of the** "Paraná de las Palmas" ; and as to the mode of proceeding **we refer the reader to Chapter** IV., "**Analysis of Observations,**" where the notations are explained and examples given how the current in feet per minute had been determined ; the position of every observation for current or sounding ascertained, and so on ; and the results may be given in tables. We only remark, that **exceptional care had been** taken in determining the surface currents across the "Paraná" ; for each a check trial had been made, which within the Palmas could not be done without impairing the results, on account of the variable nature of its currents. At the Rosario section the current does not sensibly change within a whole day ; **we may** accordingly bide our time and make any number of check observations ; **it is not material** here, **whether** the measurement of currents across the line of section will take **an hour more** or less. **On the Palmas it** may be different within a few **hours, and it was therefore of** importance to complete the trials in the shortest possible time. In calculating the velocity of current in feet per minute from the meter records, we had only considered the " Five-Minute-Observation," from which the velocity of current had been determined by the equation of the meter; and the "Minute" trials were only considered a test that no clerical error or oversight had been made, and that the instrument had been in adjustment at the time. With these premises the results of the Paraná observations are summarized in the following Tables :—

## TABLE VII.

JANUARY 24TH, 1871. LOCALITY: PARANÁ, ROSARIO SECTION.

*Soundings on Line of Section (from board the 'Aguila').*

| No. of Sounding. | Depth in Feet and Inches. | | Distance of Sounding from A of line of Base in Feet and Inches. | | Remarks. |
|---|---|---|---|---|---|
| | ft. | in. | ft. | in. | |
| No. 1 | 16 | 7 | 444 | 10 | Commenced 11 A.M. Gauge, 2 feet 5 inches. |
| " 2 | 27 | 5 | 726 | 9 | |
| " 3 | 37 | 9 | 991 | 2 | |
| " 4 | 42 | 0 | 1350 | 6 | |
| " 5 | 49 | 0 | 1718 | 1 | |
| " 6 | 53 | 8 | 2167 | 8 | |
| " 7 | 58 | 0 | 2595 | 5 | |
| " 8 | 58 | 1 | 2644 | 11 | |
| " 9 | 59 | 1 | 3000 | 0 | |
| " 10 | 69 | 6½ | 3316 | 4 | |
| " 11 | 68 | 9 | 3518 | 9 | |
| " 12 | 24 | 5 | 3865 | 4 | Changed course of steamer for intermediate points. |
| " 13 | 68 | 5 | 3461 | 4 | |
| " 14 | 70 | 11 | 3556 | 0 | |
| " 15 | 72 | 1 | 3733 | 7 | |
| " 16 | 12 | 4 | 3961 | 11 | |
| " 17 | 9 | 1 | 4177 | 6 | Terminated 1.16 P.M. Gauge, 2 feet 5 inches. On right bank at bluff. |
| C. | —6 | 0 | 4870 | 1 | |

## TABLE VIII.

JANUARY 25TH, 1871. LOCALITY: PARANÁ, ROSARIO SECTION.

*Currents across River on Line of Section.*

| No. of Trial. | Mean Time of Observation. | Current in Feet per Minute. | Distance of Locality of Observation from A of line of Base in Feet and Inches. | |
|---|---|---|---|---|
| | h. m. s. | | ft. | in. |
| No. a | 7 52 30 A.M. | 89·55 | 439 | 4 |
| " b | 8 29 30 " | 129·30 | 831 | 0 |
| " c | 10 10 30 " | 172·27 | 1238 | 4 |
| " d | 11 17 30 " | 194·95 | 1328 | 5 |
| " e | 11 50 30 " | 215·30 | 1699 | 9 |
| " f | 12 23 30 P.M. | 240·93 | 2176 | 11 |
| " g | 1 7 30 " | 255·38 | 2817 | 2 |
| " h | 1 38 30 " | 253·83 | 3559 | 7 |
| " k | 2 12 30 " | 208·42 | 3881 | 7 |

From the figures of Tables VII. and VIII. sections may be drawn representing the channel of the river, and diagrams prepared showing velocity of surface currents at various points on line of section across the Paraná. On Plate VI. the result of Tables VII. and VIII. is graphically represented without any distortion, the scale for all dimensions, horizontal and vertical, being the same; namely, 1 inch of drawing to 100 feet actual dimensions. Although the scale is a small one, the great size of the river necessitates the breaking up of the whole section into three divisions, which, joined at the lines M N of upper and middle division, and at the lines O P of the middle and the lower one, make up the whole section of the Paraná. Commencing from the line of base at A on the left bank,—an island of large dimensions of the delta and part of the province of Entre-Rios,— we have at about 33 feet the margin of the river, with a couple of inches depth of water and rushes of moderate size; the depth increasing slowly, which, at 182 feet from A, is only 3 feet 6 inches. For the first 1000 feet from A the slope of the river's bed is on the average 1 in 25; for the next 1500 feet it is on the average 1 in 75; it then remains practically level for 500 feet, the inclination being only 1 in 500. For the next 750 feet the slope continues on the average at a rate of 1 in 58; and then the first time the inclination of the bed changes to a rise, which, within the short distance of 200 feet, reduces the depth by 60 feet; from the latter point the bottom rises at a uniform incline to the right margin of the river for a distance of 1000 at a rate of about 1 in 100.

We have here a most remarkable section of a great river, in which from one bank the bottom slopes in the same direction for a distance of over 3700 feet with the regularity of railway gradients, the depths increasing from nil to 72 feet on an incline of about 1 in 50 on the average. We have nowhere met anything of the kind within a straight reach, even on a reduced scale. When the channel of a river is for a considerable distance straight and regular, it almost invariably assumes a symmetrical, generally elliptical, form. The depths increase rapidly from either bank, and then remain practically the same for intermediate points. When we speak of the outline of a channel of a river, we do not mean that of a particular section, but of a succession of sections, which should be similar in outline if the channel may be called regular, although they need not be identical. We have marked on the section each sounding according to its number in Table VII., and noted at the sounding its distance from A, on the line of base. The observed surface currents have, moreover, also been plotted on the section at their respective distance from A, and their velocity in feet per minute according to Table VIII. The direction of the currents must be taken in a horizontal sense, but in the elevation of the section they would fall into the low-water line and appear as points in projection. To make the outline of surface currents visible, they are plotted on vertical lines, and we only observe that they should be considered operating in a horizontal direction. If we unite the various points of the surface current in feet per minute at a, b, c, d, &c., the localities of observation, we obtain a curve representing the surface movement of the river from shore to shore. It is a sinuous line of double curvature, which seems to follow the outline of section as if drawn upon an exaggerated scale; and if we represent the distances which the currents traverse in a space of time equal to 13¼ seconds instead of 60 seconds, as the standard for comparison—which is quite arbitrary, and may be 1 second, or 13¼, or 60 seconds, or more—the outline of surface currents falls on the line representing the bottom of the channel of the section, and vibrates round that line, keeping close to it from margin to margin of river. The section of the

Q

Paraná on Plate VI. is, however, still too large for convenient comparison of form of the various lines; we have accordingly reduced the horizontal scale, making the inch of the diagram represent 500 feet instead of 100, thereby distorting the true outline of the section for convenience of argument and comparison, and shown on Plate V. On this section of the Paraná the similarity of outline between surface currents and the bottom of river is striking; and considering the wide range between the observed currents, varying from 90 to 255 feet per minute, the coincidence of so many points between currents and depths cannot one moment be considered accidental; it is obviously produced by a mutual dependence between the two. If we measure the distances which the surface currents at the locality of the Rosario section travel in $13\frac{1}{2}$ seconds anywhere across the line of section, these distances passed over by the current at any point are also the depths at those points; in other words, in that space of time the surface current travels over a distance equal to the depth of water at the locality. And, since the inclination of the river's surface is the same from shore to shore on the line of section, the inference is plain and warranted, that surface currents are governed by depths. Nor is there anything particular in the $13\frac{1}{2}$ seconds space of time; it obviously depends on the inclination of the river's surface at the locality. In the Uruguay, for example, during the periodical rise of the river on the 3rd February, 1871, the currents reproduced their corresponding depths in $5\frac{3}{4}$ seconds; that is, in that space of time the current travelled over a distance equal to the depth at the locality. With a greater inclination that space of time will be smaller, and may be reduced to one second, or less, as the inclination approaches more and more those of torrents. It also appears, that the law governing surface currents in the varying depths of a river's channel is independent of inclination, which only determines the extent of the movement of its waters, but not its variety on a line across its channel; for, if inclination formed part of that law, we ought to have felt its effect with the various currents and the different depths to which they correspond. No doubt the inclination was the same in each case, affecting both depth and surface current; but if it should form part of the law, the effect of the same quantity, its magnitude, would affect in a different manner the ratios of the results of the involved smaller or greater quantities; and since the result remains throughout proportional to the value of one quantity, the inference is, that the other has no direct voice in the law, and can only affect it as a constant quantity. It is important that this should be clearly understood; it is rare that we have an opportunity to fathom the laws of nature. The effective inclination at any point along the course of a river is a most troublesome quantity; it entirely escapes observation, and is always changing, and we never know what it really is. The fall between any two points along its course, which we may determine by careful levelling, is the mean inclination between those points, and not the effective one at a given locality. Here is the great difficulty which prevents us fathoming the law governing surface currents in the varying depths along the course of a river within its longitudinal section. We may have here a variety of depth with a variety of currents at the surface, but the law governing the latter is entirely obliterated by the interference of the inclination, which must change, as we shall presently show, without a possibility of measuring its amount. And, the inclination not only changes, but, by the nature of the circumstances, it alters to such an extent, that the law governing surface currents seems to be reversed, and it becomes impracticable to trace it. Suppose the discharge of a river be 1000 cubic feet per second, and that along its longitudinal section within a depth of 10 feet the surface current had been found to be 4 feet per second. If, now, within a moderate distance, say a couple of hundred yards, the depth should gradually increase to 20 feet, and all other depths in a similar proportion, the width remaining the same, we should have double the sectional area for the same discharge of 1000 cubic feet per second, and consequently the mean current would be reduced to one-half. The surface current would, in double the depth, be reduced even more than to one-half; but without complicating the argument by minor causes operating in the same direction, and assuming that the surface current would also be proportionally reduced to one-half, we must find in double the depth about half the surface current, which would lead us to believe that surface currents get reduced with an increase of depth; although precisely the opposite is the fact. The surface currents increase with an increase of depth, but since the discharge of the river, in the smaller and the greater depths producing different sectional areas, remains the same, the river so adjusts its inclination that even with the increased velocity due to increased depth the currents will pass just the 1000 cubic feet per second, and no more; and the effective inclination at the locality of double the sectional area due to double the depths will have fallen to perhaps one-sixteenth of what it had been a couple of hundred yards higher up; or, if the changes be abrupt, the effective inclination may be nil, or, for a short distance, even negative—uphill—to check the momentum of the mass as it approaches the greater sectional area, and limit the discharge to 1000 cubic feet per second. It is only for convenience of argument that we assumed the depths and sectional areas to be doubled; the increase of either, however small, will operate in the same manner, and invariably obliterate all trace of the law governing surface currents within the longitudinal section of a river by the interference of the effective inclination, which we cannot ascertain by any levelling operation.

To determine the law governing surface currents, we must therefore eliminate the interference of the change of effective inclinations by ensuring, that it had remained constant for a variety of depths; that it may no more interfere and obliterate the influence which depths may have upon currents. Across the channel of a river of regular outline and of a straight course, at right angles with the direction of the current, the effective inclination is identical from shore to shore, no matter how small or how great it may be; for, if at any point on that line the inclination were greater or smaller—no matter how little—a difference of level would at once arise below and above the line, and currents would be produced to or from shore; since, however, no such currents exist in a regular channel, there can be no difference in the inclination from shore to shore on that line. On a section at right angles with the current, all the points are therefore affected by the same inclination, and if we could but have a variety of depths, we might trace the effect on the currents. This is precisely what most sections deny, viz. variety of depth. It is not enough that a sudden depression should occur here and there, but the depression must extend for a considerable distance under the line; and besides, it must be maintained in the channel at right angles with the section; otherwise the increased depth cannot become effective, or only partially so; and, since we require that each depth should be maintained a considerable distance under the line of section, and moreover require a variety of depths each over a considerable distance, it is obvious that the channel of the river must be very wide to include all those depths, each for a considerable distance—in short, that it must be a great river; and that a small one cannot give the solution of the problem.

But even great rivers which ensure distance, usually deny variety of depths, and

Q 2

confine us to a very limited number and to a narrow range, making the extension of the rule they may indicate doubtful or vague; and if we, moreover, consider that the channel of the river must be for a great distance straight, it should be so for miles—otherwise the law would again be obliterated by another cause—it will be obvious that even here few localities can be found which will simultaneously fulfil all these conditions; and, of all the sections we had made on the great rivers, it is the Rosario section alone which disclosed and defined the law of movement of surface currents, and none of the others did so, although every one of them within its range had followed it.

## ANALYSIS OF DEPTHS AND SURFACE CURRENTS.

The geometrical analysis of the results of Tables VII. and VIII. may be briefly made by the aid of Diagram No. 1, Plate V. On this diagram the line O X may represent depths in feet, the line O Y currents in feet per minute, both commencing at O at the intersection of the lines, at which point we have therefore neither depth nor current. The scale which we may adopt either for depths or currents is arbitrary, and the choice should be so made that the variation of either should prominently affect the position of any point on the diagram. The scale for currents was taken the same for the diagram as for the sections, viz. 1 inch of diagram to represent 100 feet of currents; for depths it was, however, considerably enlarged, otherwise the points would have fallen inconveniently close together, and neither their position nor their effect could have been appreciated. The scale for depths was therefore made about seven times larger, viz. 1 inch of diagram to 15 feet of actual depth. We may now from the section of the Paraná determine the depth corresponding to the locality where the current was observed; at some currents a sounding happens to be very close; at others, the locality falls between two soundings, and the depth may be determined from the outline of section. For each observed current called a, b, c, d, . . . . in Table VIII., we obtain from the section the corresponding depth, viz. 18 feet 9 inches, 31 feet 7 inches, 40 feet 10 inches, 42 feet 0 inches, &c., . . . . which may be plotted on the diagram in the following manner:—From O on the line O X a distance of 18 feet 9 inches having been marked off by the horizontal scale of the diagram, at that point a vertical line parallel to O Y is drawn, marking on the latter by the scale for currents a distance of 89 feet 6 inches, as the velocity of surface current per minute, which will determine a point marked a on the diagram. In a similar manner the depth of current b having been marked off on O X from O, and on a line at right angles parallel to O Y, the distance travelled over by the current in one minute having also been determined from the line O X, a point is obtained marked b on the diagram; and so on with every succeeding current observation of Table VIII. we obtain a series of points spread over the diagram called a, b, c, d, e, f, g, h, and k, all of which belong to the observations of the Rosario section; for the present other points on the diagram without letters need not be taken into consideration. If we now take point f, about midway the section, not far from the greatest depth, and near which we have a sounding marked No. 6, and unite f with O by a straight line called O A on the diagram, we find that all the points a, b, c, . . . . &c., as determined by the velocity of the currents, except one, fall close to the line O A, and vibrate to the right and to the left of it; and that the line O A would be a mean line through those points, substantially representing the whole series. It should be remembered that O also belongs to the system, since in a depth of zero the current would be nil; we have accordingly a number

of points spread over a distance from nil to 60 feet with considerable uniformity and without long intervening spaces, all of which substantially determine the law, *viz.* that the
surface currents increase as the ordinates of a straight line originating from the intersection
of the co-ordinates, the abscissæ representing the depths corresponding to those ordinates.
That the experimented points, a, b, c, &c., . . . . do not exactly fall upon the line itself, in no
manner calls the above law in question, if we consider what the soundings and the currents
necessarily represent in the diagram. The soundings determine a specific section, which,
however regular the channel of the river may appear, can never be identical in succession;
there must be some variation in the depths on every longitudinal section along the line of
the various currents; and these will not be governed by one particular section, but by the
mean of a number of successive ones preceding perhaps a mile the trial section; the surface
currents will adjust themselves to the mean depth over which they pass, and that mean
necessarily differs a little from the trial section; and, if the channel be of a regular outline,
the differences will be in succession positive, nil, or negative, producing substantially the
same sectional area. It would be mere chance that the trial section should happen to be in
outline the mean of a number of preceding ones, and only in that case could all the points
of the observed surface currents fall into the line O A. One of the most striking illustra-
tions that surface currents, with similar inclinations of a river's surface, are only governed
by depths, afford the first and last current observations on the Rosario section; and it is
fortunate that the last trial fell between two soundings close to one another, otherwise, if
we should not possess soundings Nos. 12 and 16, it would appear greatly to deviate from
the rule on account of the rapid change of depths in that part of the river. These two
current trials are about 3400 feet apart, not far from opposite shores; they happen to be in
depths only a couple of feet different, and after the current had steadily increased with the
greater depths for all the succeeding points, it had suddenly dropped with the decreased
depth at the last observation to nearly the lowest velocity, falling close to line O A to a
position nearly corresponding to the depth at the locality. When the current observation
k was made, as with every one of the others, we had no idea what the depths may
have been below the keel of the little boat, moored at both ends on line of section for current
trials; and no attempt was then made to take a sounding in the strong currents of the
Paraná. Observation h deviates from the rule, the velocity of current falling about
60 feet short of that due to depth; but the exception proves the rule if the locality be
exceptional; it happens to be so with h, being not far from an abrupt rise in the bed
close to the rocky shore; and by the form of the depression from sounding No. 9 to No. 12,
and the above circumstances, the inference is, that the depression is local and not general
in the longitudinal section in that reach of the river.

Although the Uruguay does not form the subject of this chapter, we may advantageously
consider the analysis of its depths and surface currents represented in Diagram No. 2, the
result of which had also been transferred to Diagram No. 1 by a series of points without
letters close to line O B. The scale for depths and currents is the same in both diagrams,
and they were obtained by similar proceedings. The section of the Uruguay is very fine,
the river's course is for miles within a straight channel confined on both sides between high
banks, and with a width of half a mile—yet, within this great width the range of depths,
maintained for a considerable distance, is limited to between 20 and 30 feet; we have only
one observation between 0 and 20 feet, whilst we have six between 20 and 30 feet. Here,

again, all the points except one fall close to the line O B, and oscillate to the right and left of that line ; the exceptional point is in this section also close to shore at abrupt changes on the rocky side of the river ; and it is clear that the depth, which corresponds to that current as obtained from the section, is local and not general in that reach of the river, considered in a longitudinal direction. The Uruguay, with a much greater inclination of its surface, with a fall of about $5\frac{1}{2}$ times that of the Paraná, follows the same law determined by the Rosario section ; but it would not have been safe to deduce it from the Salto section of the Uruguay, because the straight line O B, though it divides the points between 20 and 33 feet more favourably than any curve of the second or higher order, would nevertheless leave it a matter of opinion whether a parabola U, for example, passing through O, and the velocity point of current corresponding to 30 feet on line O B, would not also symmetrically divide the points if we leave the solitary current at $11\frac{1}{2}$ feet depth out of consideration ; and, should the parabola represent the law of increase of surface currents, then the latter would increase not as the depths but as the square roots of the depths.

The Rosario section, however, defines the question, and does not leave it doubtful or a matter of opinion. If we pass a parabola through O and f, marked P on Diagram No. 1, we find that with the exception of points e and g, which come near the parabola because they are near f, all the other points get not only farther and farther from it, but they all occur on one side, and do not oscillate to the right and left of the curve. We might lay a parabola through an intermediate point like e, but it would only divide the points in two systems, one of which would be above, the other below the curve, and they would not alternate; or, if we produce a parabola through a, marked R on the diagram, two points would come near it; not only would, however, all the others fall above the branch, but they would differ more and more as the depths increase. Nor would parabolas passed through any point of the system taking O Y as the axis of the curve give better results. We have shown two such parabolas, one through a, the other through f, marked respectively T and S on the diagram, and these deviate even more rapidly from the various points than the curves referred to with O X as their axis. We are therefore led to the conclusion, that surface currents with a given fall do not increase either as the square or as the square root of the depths, but as a linear function in direct proportion to the depths. And this will further demonstrate, that all equations in which the surface current is determined by an algebraic expression and in which depths appear under the square root, or under any power except the unit, are erroneous and defective ; and although such formulæ might be so adapted by the introduction of coefficients and constants as to give a correct result for a certain depth, and within a narrow range come near enough for ordinary purposes, they must be erroneous to a serious extent whenever the variation in depths happens to be considerable, or whenever they deviate from the experimental depth for which the formulæ had been determined. Suppose a formula were so constructed that it agreed with the result of observation of point a, and, for a depth of 18 feet 9 inches, and the fall per mile ascertained at the section, would give a surface velocity of 89 feet 6 inches per minute ; and suppose that in the formula so constructed the depth appeared under the square root, we should have for the same fall and varying depths results which would come within the parabola O R of the diagram, and that curve would mathematically represent the increase of surface velocity due to increase of depth at the same fall per mile. Close to point a, say within 5 feet more or less, the deviation of the result of the formula from the actual surface velocities due to those depths would be immaterial; but when we begin to double the depths or to halve them, we find, that the result of the formula must fall short of the actual surface current in the former by 40, in the latter it would be in excess of the true current by about 50, per cent.; and, if we treble the depths or reduce them to ½rd, the result of the formula would in the former fall short about 70, in the latter it would be in excess about 100, per cent. of the actual surface current; though the same formula would give correct results for a depth of 20 feet or thereabout.

It might be thought that the object of most of our formulæ is to determine the mean and not a particular surface current, and that our formulæ do not deal with particular depths, but only with mean depths, called hydraulic-mean-depth of river. We have already had occasion to note in a former chapter the tendency of generalizing matters when we cannot deal with them in detail; it is more convenient to deal with mean currents and mean depths, because they are imaginary quantities, and their accuracy of conception and treatment cannot be directly contradicted by observed facts. As to the possibly mysterious effects of hydraulic-mean-depth, which appear in our formulæ, instead of a definite sounding obtained by measurement—we refer to the middle division of the section of the Paraná, Plate VI., and ask, whether at any point of that division the direct sounding giving depth is not also the hydraulic-mean-depth of river at that locality? whether to the right and left of any of the current observations on that division, equal to the widths of most of our European rivers, the depth of the current observation is not precisely the hydraulic-mean-depth of river as understood by our best authors? and we may also include the lowest division; indeed, from the base at A, to a distance of 3200 feet, every depth at the current observations is the hydraulic-mean-depth of river at that locality for that current; and it should be remembered that the whole of our arguments are based upon the depths and currents of that part of the river's channel. There is therefore nothing mysterious in the operation of a term called "hydraulic-mean-depth" of river, which, under favourable circumstances of channel, represents the depth at the locality of the current; and under unfavourable circumstances, it is an approximation to that depth, had the channel not been irregular. We may now consider whether our formulæ will have a better chance of ensuring correct results by dealing with mean instead of surface currents as long as the depth, whether as a "hydraulic mean," or as a direct sounding at the locality of the current, should remain under the square root. The observations on the LA PLATA and the PALMAS have conclusively shown that the mean current increases more rapidly than the surface current, no matter whether the increase be due to greater fall or to greater depth, and we have direct trials that the mean current is particularly affected by depth. See p. 45, Table III., trials No. 22 of 29th, and No. 19 of 30th December, 1870, and subsequent analysis of mean and surface currents—that for the same surface currents the mean was greatly increased on account of depth alone, that consequently the mean current increases more rapidly with an increase of depth than the surface current. Bearing the above in mind, it follows, that if our formulæ erred 70 per cent. from the true surface current as soon as we trebled the depth for which they had been adjusted, they must err more if their results are to represent mean instead of surface currents; and we believe this to be a serious matter, of consequence to hydraulic engineering;—deserving the utmost attention and reflection. It is not our wish to supersede the formulæ of others. The opportunities we have had to trace and to ascertain the laws which determine the movement of water within natural channels, had not been shared by authors of ability; we place accordingly before them the result of our observations for their consideration.

## ANALYSIS OF INCLINATION.

By the aid of Diagram No. 1, Plate V., we may not only check some of our former conclusions on the result of observations on the La Plata and the Palmas by the light which the Rosario section sheds upon currents, depths, and inclination, but we may also determine by the same diagram the inclination of the Paraná at Rosario. Before we, however, can do so, we must consider the La Plata and Palmas trials by the results of the Rosario section.

The line O A of the diagram is a velocity line. We know that the inclination or fall of a river's surface will approximately increase or decrease as the square of the corresponding velocity. The angle A O X will, at the same depth, increase or decrease with the velocity, and will therefore also indicate the greater or the smaller fall at the locality. A series of observations were made on the La Plata, December 30th, 1870, from 3.30 to 5.30 P.M., as represented on Diagrams 1, 2, 3, Plate II., and considered pp. 38 to 43, Table I. On Diagram No. 2, Plate II., the curve representing surface currents during trials is shown, and we have a minimum, mean, and maximum surface current, which from the diagram appear at 106 feet 6 inches, 111 feet 8 inches, 116 feet 9 inches per minute; the depths at the locality, determined by the tide records, were 24 feet 6 inches, 24 feet 2 inches, and 23 feet 10 inches respectively. If we now plot these depths on Diagram No. 1, Plate V., and their corresponding surface velocities in the manner adopted for points a, b, c, . . . . . &c., we obtain three points marked on the diagram ⌀ which belong to the La Plata. Through each point a line from O may be drawn; we have, however, only shown the prolongation of these lines at the margin of the diagram to avoid the crowding together of too many lines, which would interfere with clearness. For these Plata points we have, moreover, the inclination of the surface determined from tide records, see pp. 53 and 54; the fall for the minimum, mean, and maximum surface currents was found to be 0.342, 0.388, 0.444 inch per mile, and we find the velocity line of the Paraná near Rosario enclosed by the velocity lines of the La Plata between the minimum and the mean current; for which the corresponding inclinations are known. We accordingly find, that the inclination of Paraná's surface near Rosario and that of the La Plata in the outer roads from flood to ebb is the same; between the limits of 0.34 and 0.39 inch, and we may take it at 0.37 inch per mile. The La Plata results deserve confidence, because the bed of the channel is very uniform, and there is nothing to disturb the movement of the currents; the differences in depths were obtained from the gauge records and not by soundings, and are therefore correct to the inch. We must bear in mind, that we are now tracing microscopical quantities which would escape the most refined levelling instruments in experienced and good hands. It is a question of the value of the figure in the second decimal of the inch per mile; and, since we can trace it by the aid of surface currents and the law which governs their movement, it is evidence of its extreme delicacy, and another point in favour of its truthfulness and accuracy.

The Palmas depths and currents might be analyzed in a similar manner, but we have already observed, page 91, when the outline of its surface currents across its line of section was under consideration, that it was significant and exceptional; and that the cause of disturbance would be traced farther on. We might prepare a diagram for the Palmas section referring to depths and surface currents, but we should find it confused, vague, and irregular. If we look at the outline of the Palmas surface currents across line of section Plate IV. at "Plan of Survey," and compare that outline with the exaggerated bottom of rivers shown by a dotted line, we perceive a distorted similarity between the two; the currents still increase with the greater depths, but the rate of increase is different as we approach either from the right or left bank into the river. All the currents from the right shore towards the Observatory fall short of velocity in comparison to depths,—those from the left bank towards the middle of river are in excess, and there seems to be a crowding of currents towards the left shore of the Palmas. From the sketch of the locality in the Survey Book it appears that the channel of the Palmas is throughout on a gentle curve, the left bank being on the concave and the right on the convex side; and that about one mile above the section the bend is getting sharper in the same direction; the radius of curvature might be from one to two miles. The river encroaches on the left bank; this is also shown by the form of the section, the right shore being a marsh frequently submerged; the left bank is about 3 feet above water level. We trace here other important matters. A slight curvature in the channel of a river greatly confuses the currents, and drives them from the locality which previously had determined their velocity. Centrifugal force is here a cause of disturbance. This force increases as the square of the velocity; and consequently, although every current across the line of section is affected by it, those of double and treble velocity are four and nine times more affected respectively; and, by the greater pressure which these exert, they are carried towards the concave side, and they constantly mix up with and displace the weaker currents. None of the surface velocities of the Palmas section can therefore precisely follow the law between "depth and velocity," and all must deviate; those near shore will deviate most and those about midway least. Here is another illustration of the difficulties which often surround hydraulic questions. Not only had the constant change of inclination of a river's surface obliterated and frequently reversed the operation of the law between "depths and currents," but apart from that cause of disturbance, the law may again be seriously affected and even obliterated by another circumstance, a slight bend in the river's course. We may nevertheless take a surface current of the Palmas for which we know the fall per mile and determine its position on Diagram No. 1, Plate V., by a similar process as applied to the other points. The inclination of the Palmas surface (see the Table, page 81) is known for certain hours, and we may take No. 2, which near 6 P.M., January 18th, was 0·411 inch per mile; the surface current having been 209 feet per minute at 5 P.M. at the Observatory, within a depth of 49 feet 7 inches of water. If these figures be plotted on Diagram No. 1, a point is obtained for the Palmas which comes close to the La Plata and Paraná lines, showing, that even with the disturbance of curvature that point gives the inclination of the Palmas substantially to be the same as that of the La Plata or Paraná near Rosario. We know, however, that the Palmas point falls short of its true position by about 26 feet per minute, because it ought to fall a little above the Plata line marked 0·39 inch per mile; and this circumstance shows, that to a certain extent the current was mixed up with others of smaller velocity, which reduced it from about 235 to 209 feet per minute; assuming, that the depth of 49 feet 7 inches was an average in a longitudinal direction of the channel. It should be,

R

however, remembered, that even with the disturbance arising from curvature, the inclination of the surface of the Palmas as determined by the current and the corresponding depth is not $\frac{1}{10}$th inch short of the true fall per mile; a quantity which would still escape the best levels.

It might be thought, that if surface currents are so easily disturbed by irregularities in the channel of a river and through bends in its course, the rule between "depths and currents" may not be of much practical value, although it may be of scientific interest. We consider it of paramount importance that, above all, the causes should be clearly understood which determine the movement of water within channels; that we should know what happens as a matter of fact, and not have to rely on ingenious theories as matters of hypothesis. If nothing else were gained, we should have acquired substantial information. It has already been shown how the discovery of the true movements of currents and the causes which govern them affect our formulæ and the result of our calculations. Nor is this all. Knowing the operation of each particular cause, the joint effect of which produces the river's movement, we are enabled to foresee and to foretel what must happen under different circumstances, and this is of the highest importance; and its successful practice proof of the greatest accomplishment of an engineer. The volume of a river or the velocity of any of its currents at a locality, may be determined with ease and accuracy by means of a good meter and the current integrator, without our possessing any knowledge whatever as to the causes which produced those movements; it is a mechanical and a geometrical operation, like surveys generally, their object being the accurate expression of certain facts; the comprehension and the clear understanding of the causes which produced those facts, embody the thought and the ideas of the profession; and these are of primary importance.

Although the rule defining "depths and currents" is so easily disturbed by irregularities or bends of channels, it is nevertheless, even under unfavourable circumstances, a much superior guide than any argument based upon assumptions. Depths govern currents. It is only by their relative value that we can form an estimate of the effective inclination of a river's surface at a given locality, and we know of no other means which, in many cases, could even indicate it. Levelling will approximately, and sometimes correctly, determine the inclination of a river's surface at a given locality if the fall be great and the channel not irregular, because a short distance will disclose the fall with a powerful instrument with sufficient accuracy for ordinary purposes; the mean fall within a short distance may then be taken as the fall at the locality. The current corresponding to a certain depth is, however, a much more delicate and certain measure than any levelling. Anywhere within a straight reach and a regular channel we need observe only one current and the corresponding depth, and the inclination is determined. If the depth at the locality happen to be maintained for a moderate distance across the river and along its course, the fall will be found with an accuracy rarely required for ordinary purposes. The inclination may under such circumstances be determined by one observation without any section across the river; and, since it may easily be repeated along its course in situations where its channel had enlarged or contracted, we have a ready means to determine the "gradients" along its line. We must, however, remember that the observation does not give the fall directly; it only reduces the varying velocities to one common standard. We need but divide the surface velocity by the depth in each case, and if the result be the same, the fall will be identical;

if the various quotients differ, so will the fall differ in each case; and we know at once that in any two rivers, however different in size, where the surface velocity divided by the corresponding depth, will give similar results, the inclination will be the same. Without the aid of formulæ, the fall may at each locality be ascertained as follows:—We may, for ordinary purposes, take the fall proportional to the square of velocity; mathematically considered this is not correct, but it is a good approximation and near enough for practical requirements. The velocities having been observed along the course of a river, especially at localities where its width and depth should change considerably, and the quotients determined and each squared, the new figures may be considered as proportionate values of the fall; and, by an arbitrary scale they may be plotted for the distances representing the successive stations where observations had been made. Commencing at the first point, the distance to the first station may be plotted on a horizontal line by a scale for horizontal dimensions; at the end of which the value obtained for the proportionate fall may be represented by another arbitrary yet convenient scale for vertical dimensions, and the new point connected with the first will represent the relative inclination; from the second point another horizontal line should be drawn, and the next distance marked, at the end of which the proportional fall having been plotted, the relative inclination within the next reach will be determined; and so on with the various stations where observations had been made. A succession of inclines will be thus obtained connecting the first point with the last; and the value of the proportional figures representing the fall, now remains to be ascertained. For this purpose, a level should be run from the first to the last point, and the actual fall between the terminal points having been found, it will represent the difference of level on the drawing between the first and last point. The value of the arbitrary vertical scale may be accordingly ascertained, and the fall between each station may now be expressed in feet and inches. By this proceeding we obtain with accuracy the various "gradients" along the course of a river without the use of any formula. Although the gradients might easily be calculated by the aid of a correct formula, it is desirable that a check should be obtained, and the difference of level between the terminal points should be always ascertained as a check on the various results. Without a correct knowledge of the gradients of a river, its improvement is in a great measure guesswork; and good intentions may result in serious injury.

The observations on the Paraná also enable us to deal with rivers of irregular channel, where a single observation of surface current and corresponding depth would be of doubtful value. We have seen that surface currents are governed by depths, and that the former always represent the mean depths which for a distance preceded the locality on the line of current, provided the channel be straight. Several soundings ought therefore to be taken on the line of current above the locality of trial, for a distance which will depend on circumstances, and which will be greater with a greater depth; the mean of these soundings will determine the effective depth for the surface current. Since, however, the irregularities also appear on the cross section, we may, as their measure, take the mean surface current across line of section, which will be governed by the mean effective depth; also called hydraulic-mean-depth. The result will be usually correct, however irregular the bottom may be; because, these will be positive and negative in succession, and substantially eliminate one another. We find, for example, the mean depth of the Rosario section 38 feet 7 inches, and the mean current within the surface movement limited by the rushes equal to

166 feet 6 inches per minute ; and if we plot these figures on Diagram No. 1, Plate V., we obtain a point marked "Paraná mean," close to the line of velocity, of which it falls a little short, for reasons which will be explained farther on. In a similar manner we find the mean depth of the Uruguay, at the Salto section, to be 26 feet 11 inches, and the mean surface current from shore to shore 285 feet 8 inches, which figures plotted on the diagram give a point marked "Uruguay mean," which falls precisely on the velocity line of the Uruguay, for reasons we shall offer farther on ; and we only remark, that the velocity lines of the Paraná and Uruguay had been determined previously and independently of the mean points, and that their coincidence or variation is explained by circumstances which shall be specially considered.

Accordingly, with an irregular channel of river, a cross section should be made at the locality, and the surface currents from shore to shore observed ; and the mean depth of river taken for the mean surface current, which will ensure satisfactory results. The section should be at the end of a straight reach of at least a couple of hundred yards length, within which the currents may have adjusted themselves to the form of channel before they reach the locality of observation.

We do not enter into the consideration or construction of formulæ for the correct expression of the rules derived from our observations on the great rivers. We are at present engaged mainly with an historical statement of facts collected during our surveys, and the remarks we offer are given with a view to explain circumstances and details for the better comprehension of those results. To enter into a philosophic consideration of the various matters which the trials placed on record, and to embody these in correct mathematical expressions for a variety of purposes with which the science of hydraulic engineering has to deal, would, we believe, not be in harmony with our object, and require a special work of considerable extent, and demand special treatment. Those who are familiar with the thought and expression of mathematics will easily combine the results and define them by correct formulæ ; and others, whose occupation will not admit constant practice with theoretical considerations, will be equally guided by the results of our observations without the inconvenience attending the frequent use of expressions, which, in a measure, might be foreign and troublesome without a separate and specific treatment of the subject.

## CENTRES OF GRAVITY OF SECTION AND CURRENTS.

Another interesting and instructive investigation is the relative position of the centres of gravity of the section and the currents of a river. By centre of gravity a point of a surface or of a mass is understood, which, when supported, will balance all the surrounding points in any position we may place the surface or the mass. We may in effect imagine all the points condensed into that one, the position of which is definite for a given surface or mass of any form, however irregular it may be. With regular surfaces or masses it may be geometrically or mathematically determined ; but with all, whether regular or irregular, it may be promptly and accurately ascertained by experiment. We need but suspend the surface or the mass from two different arbitrary points, and the intersection of the vertical lines from the points of suspension will determine the centre of gravity of the surface or mass under consideration. The experimental mode was followed to ascertain the centres of gravity of the irregular surfaces representing section of rivers and area of superficial currents, and the result is correct within the puncture of the paper. The surfaces were traced and transferred on drawing-paper, their outline was cut out and suspended, and the vertical lines drawn; the intersection of lines giving the centre of gravity, a few check trials being readily obtained. By this proceeding the centre of gravity of the Rosario section was found, and shown Plate V., marked " centre of gravity of section," and in a similar manner for the surface currents another point was found, marked " centre of gravity of currents." The position of the centre for section is unchangeable; whilst that for the currents may occupy any position upon a vertical line, according as the measurement of currents be made by minutes or seconds; if, for example, the currents in the Paraná had been represented for a space of time equal to $13\frac{1}{4}$ seconds, the centre of gravity of currents would fall on the same level as that of the section. What is of importance is the horizontal distance between the two centres, which in the Paraná, though the points come near each other, still amounts to about 17 feet, the centre for the currents falling a little towards the right bank of the river. With the Uruguay the horizontal distance between the two centres is hardly perceptible; the two lines nearly cover one another, and the slight difference may come within the possible accuracy of construction. Such coincidences between the centres of gravity of section and currents cannot one moment be considered accidental; they appear to be a summary of the law between depths and currents; and they confirm substantially the arguments which we followed in detail with each point in succession, analyzing the dependence between depths and surface currents of the Rosario section. If no such dependence existed, the centre of gravity of currents might be anywhere in reference to that of the section, and we shall see that even the slight deviation in the Paraná is accounted for. It should be further considered that the outline of surface currents on the Paraná, as well as on the Uruguay, is very irregular; a number of bends and projections in succession, sharp and flat; and there is no similarity between the outline of currents of the Paraná and the Uruguay; yet, these dissimilar surfaces show the closest agreement with their respective sections; and, the natural inference is, that the currents as a whole, by their centre of gravity, follow the section of river as a whole represented by its centre of gravity, in the same manner as each individual current was traced to follow its corresponding depth. The centre of gravity of section represents the sum of effect of all the various depths condensed in one vertical line passing through that point; the centre of currents represents precisely the same thing in reference to currents on another vertical line. These two lines fall substantially together: the conclusion is, that the one is the expression of the other, and that they represent the same thing interpreted by different mediums; that depth and velocity are interchangeable quantities.

In the Uruguay we have, however, not only a perfect agreement between the centres of gravity of section and currents; but if we determine the mean depth of river and the mean velocity of superficial currents from the respective areas which produced those centres, and plot for the mean depth the mean current on Diagrams Nos. 1 and 2, Plate V., the velocity point called " Uruguay mean " falls precisely upon the Uruguay velocity line, determined previously and independently of the several Uruguay observations. The sectional area of the Uruguay was, on the 2nd and 3rd February, 1871, in round figures, 71,200 square feet, over a width of river from margin to margin of 2641 feet, giving a mean depth equal to 26 feet 11 inches over that width. The area traversed by the currents in one minute was 755,000 square feet, giving over the same width of river a mean distance for currents of 285 feet 8 inches per minute ; and, these figures plotted, the point called "Uruguay mean" falls into the velocity line of the diagram. It should not be thought that this will follow as a matter of course ; that the centres of gravity falling in the same line on the section will make the "Uruguay mean" point also fall on the velocity line of Diagrams Nos. 1 and 2,—because we should remember, that we might actually reverse the section, and make that part on the right appear on the left ; and, leaving the position of currents the same, the centres of gravity would nevertheless fall on the same line ; but inextricable confusion would be produced as to the law between depth and currents ; and in plotting the velocities for the new depths the line would appear an S-shaped curve, never even approaching the origin O of the co-ordinates.

It is the close agreement of the Uruguay quantities which called specially our attention to the slight variation of those of the Paraná, which appeared in the section as well as in the diagram when centres of gravity were compared, or the position of mean point in reference to velocity line was considered. Had the variation been confined to either section or diagram, it might possibly have arisen from some inaccuracy ; but its simultaneous appearance in both, led us to suspect a cause of disturbance. We then referred to the Palmas section, and found that the position of the centres of gravity of section and currents here differed largely, the horizontal distance between the two being 46 feet 6 inches ; and, since the width of the Palmas is only about one-fourth of the Paraná at Rosario, the variation in the Palmas is eleven times as much as that of the Paraná. Having also determined the mean depth of the Palmas section (which happens to be the same as that of the Paraná near Rosario), and the corresponding mean surface current, and having plotted their values, it was found that the "Palmas mean" deviated from the velocity line then in force (1.15 P.M., January 19th, when the surface current across section had been measured) by 19 feet per minute, whilst that of the "Paraná mean" only fell about 6 feet short of its velocity line. It should be remembered that comparisons with the Palmas are complicated by the constant change of its inclination and the velocity of its currents, on account of the interference of the Plata tides, all of which make the Palmas velocity line variable, not like the lines of the Paraná and Uruguay, which are invariable as long as their volume remains the same. We may, nevertheless, make comparisons at certain hours of the day ; it was found that the "Palmas mean" falls considerably short of its corresponding velocity line, and that the centres of gravity of section and currents are considerably thrown apart. We know that in the Palmas the disturbance of position of currents had been caused by the bend of the river at the locality of the section, see p. 121, and we have now a measure of that disturbance by the displacement of centre of gravity of currents and that of section, which at that time amounted to 46 feet 6 inches. We also note, that the centre of gravity of currents is thrown towards the concave side of the river's bend, and that the mean surface current of 125 feet 8 inches plotted on the mean depth of river of 38 feet 9 inches, falls a considerable distance short of the velocity line then (1.15 P.M., January 19, 1871) in force ; the latter, for a depth of 50 feet, having been determined by a current of 185 feet 6 inches per minute.

The inference is, that since these marked deviations had been caused by a bend in the course of the Palmas, so the disturbance of centres of gravity and of the "Paraná mean" of the Rosario section may be due to the same cause ; and that, after all, the channel of the

**Paraná at the** Rosario section is not straight, but that it would be found on a gentle **curve.** We thereupon searched the Survey Books, and among a variety of notations we found **the** following entries :—

"Base, 3000 feet on left bank and near margin of river; magnetic bearing of base looking up river, 305° 30'."
"Direction of right bank looking up river, 311° by prismatic compass."

It follows, that since the base was on the left bank, and its ends are within **40 to 50 feet** from the margin of the Paraná, the right shore by its magnetic bearing **inclined 7° 30'** with the line of base, or with that of the left bank looking up river; and, since **the bluff** sighted by the compass was, according to survey, **close** upon three miles **from the section,** the angle which that sight formed **with the section was,** instead of **90° 0',** as on the left bank, only **82° 30';** **which circumstance determines the radius of curvature of right** bank approximately at about 12 miles, the centre of radius being on the left bank. The right shore forms accordingly the concave side of the river; and if its submerged channel follows the curvature of the bluff above water, we should then expect the centre of gravity of currents thrown from that of the section towards the right bank. If we consult the section we find, that the centre of gravity of currents falls 17 feet 6 inches from that of the section towards the right bank, which is strong evidence that the course of the Paraná is on a slight bend at that locality. We further find that the "Paraná mean" falls a little short of the velocity line of river, just as the Palmas did on a much greater scale on account of the sharper bend, and this is additional evidence, if not conclusive proof, that the Paraná is also on a bend. If we, however, further consider that according to the plan of survey the right bank of the Uruguay, at the Salto section, is for three miles on a straight line, and that with this section the centres of gravity fall practically on the same line, and the "Uruguay mean" also on its velocity line, there need be little if any doubt on the point, that the disturbance indicated by a slight displacement of centres of gravity and mean point is, at the Rosario section, caused by a slight curvature in the channel of the Paraná, and that but for this circumstance the centres of gravity would have fallen in the same vertical line, and the "Paraná mean" would have come exactly on the velocity line of the river, as determined by Diagram No. 1.

We have here an illustration of the soundness of the rule "never to improve an observation" to make it agree with accepted views and ideas. Slight differences and variations may be easily obliterated by "improving" one or the other of the inconvenient observations, and so bring it in harmony with the rest; but we should **only effectually shut** out the light which might have led to important discoveries. **Not only have the observa-**tions on the Paraná, patiently considered, led to **the discovery of the dependence between** depths and surface currents defining their operation, **but we have been enabled, from the** same observations, to trace with equal success causes of disturbance.

The relative position of the centres of gravity of section and surface currents in a horizontal direction will therefore at once determine the question whether the channel of a river be straight or bent at the locality under consideration, even if the banks were invisible, or the submerged channel should not follow their outline above water; the centres also show on which side the concave part of the channel lies. It appears, moreover, that a gentle bend will considerably disturb all currents, as shown on the Palmas where the radius of curvature is about 1½ mile; and that a bend in the course of a river causes a much more serious disturbance with all currents, than irregularities in its bed. Nor is it matter for surprise that it should be so. The irregularities on the bed are caused by local variation of depth; and, in their effect these will be, in succession, positive and negative; some increasing, others reducing the corresponding velocity of current. The effect of a bend on the currents of a river is, however, all in one direction, carrying the stronger currents from the centre of curvature towards the concave shore; and thus, during the continuance of the bend, constantly mixing up the stronger and the weaker ones, tending to produce a mean surface current; also tending to make currents from surface to bottom more and more alike. After a succession of sharp bends upon a sinuous channel of a river, we may expect the currents "meaned," and to be much alike wherever they may be measured; the operation of the law between depths and currents having been modified and perhaps obliterated. Whenever trials for surface currents are desirable, to ascertain the effective fall at a locality, or trials for other currents are intended, the first consideration should be a straight channel above the locality of observation for a considerable distance; the second that the width should be uniform; and lastly, that the depths should not greatly differ at similar distances from shore. How long the straight reach should be, to ensure the currents having adjusted themselves to the special form and depth of channel, had not yet been determined from a sufficient number of observations. We are inclined to think that the distance will be governed by the maximum depth; and that with ordinary velocities of current of from two to three hundred feet per minute it may be about 100 to 150 times the maximum depth at the locality. The position of the centres of gravity of section and of currents will show how far the locality had been favourable for trials and its selection good, and how far the results may be relied on for absolute and comparative values and for argument.

## ANALYSIS OF SECTION.

In the preceding divisions of this chapter, and also in some of the former, frequent allusion has been made to the Rosario section of the Paraná; its outline and its peculiarities have been considered and analyzed, and under the present heading we can but recapitulate, in a condensed form, our observations on this great section.

The channel of the Paraná at the locality of observations is for miles practically straight; it was believed to be straight until the position of centres of gravity of section and currents disclosed it to be on a gentle curve, the radius of curvature being about 12 miles; the right bank forming the concave part of the river's bed. Referring to Plates V. and VI., a base of 3000 feet had been measured on the left bank, close to the margin of water upon a sandy beach, which insensibly slopes into the river's bed from a depth of a few inches to 60 feet at a very uniform rate for about 3000 feet from the left margin; for the next 500 feet the slope is not so regular, yet always maintained in the same direction, when at 3753 feet from the base the maximum depth of 72 feet is reached, from which point the bottom rises abruptly to within 12 feet in a distance of 200 feet. This is evidently a submerged rocky shore, probably similar to that above water on the right bank, an almost vertical rocky bluff of the "tosca" formation. We had devoted some attention to this formation, and we believe it to be the remains of an ancient coral-reef of the Porites and Madrepore species. At a distance of 4820 feet from the base the right margin of river is reached; the clear width of the Paraná from margin to margin of water on line of section being 4786 feet 7 inches on the 24th January, 1871, at a level corresponding to ordinary low water. The sectional area of the river corresponding to low water was found to be 184,858 square feet, giving a nominal mean depth of 38 feet 7 inches from margin to margin of water, over a distance of 4786 feet 7 inches. The effective mean depth is, however, much greater, because the bulk of the river's volume—almost the whole, viz. 99½ per cent.—is conveyed within 3865 feet from the base; and, deducting the distance from base to margin of rushes on the left bank, equal to 142 feet, the effective width of river for 99½ per cent. of its volume is 3723 feet, which in reference to the area enclosed gives an effective mean depth of 47 feet 6 inches; instead of 38 feet 7 inches as the nominal one. The centre of gravity of section is at a distance of 2407 feet 6 inches from A on the line of base, and 27 feet 2 inches below the surface of the water. From soundings No. 10 to No. 15, especially in the neighbourhood of soundings No. 14 and 15, the depression in the river's bed seems to be local and not general, if considered in a longitudinal direction at right angles to the line of section, because the current h at 3539 feet from A falls considerably short of velocity in reference to the depth at the locality. The proximity of the abrupt rise in the bed may and will have its effect on the currents, independently of that of the bottom 70 feet below the observation; but it is probable that the depression, amounting to about 10 feet over a distance of 500 feet, is nevertheless not general, and not maintained for one mile in a longitudinal direction of the channel.

The currents increase steadily and at a uniform rate from the left bank of river, from nil to a velocity of 272 feet per minute, at a distance of 3316 feet from the base, within a depth of 69 feet 7 inches; as determined by the curve representing surface velocity across river on line of section. The current then gradually decreases; and between observation h and k the decrease is rapid, from 254 to 108 feet per minute, the rapid change of velocity occurring at the same locality as the abrupt change of rise in the bottom of river from a depth of 70 to 20 feet. The area passed over by the surface currents from bank to bank in one minute was, on the 25th January, 1871, at the locality of the section, equal to 766,000 square feet; which over a width of 4786 feet 7 inches would give a nominal mean movement of current of 160 feet 0 inch per minute; but if the distance be taken into account over which the actual movement is limited, as determined by the rushes on the left margin and boulders on the right, and which would limit it to about 4600 feet, the effective mean movement would be 166 feet 6 inches per minute, representing the mean surface current. The centre of gravity of currents is 2425 feet 0 inch from A, and the difference between the centres of gravity of currents and section is accordingly 17 feet 6 inches in a horizontal direction, that of currents falling between the right shore and the other centre.

At ordinary flood level the sectional area of the Paraná at the locality of the Rosario section would be between the rocky bluff and the base 243,000 square feet, but the left bank would be flooded for many miles; and though the current over the flooded islands would be small, a considerable volume would nevertheless pass over the great additional sectional area. The currents had not been observed at flood level; but their velocity may be approximately determined by calculation; first, due to increased depth; readily and accurately ascertained by the law between depth and surface currents for the same inclination; and second, due to increase of inclination, which would be about 12 feet for a distance

s

of 170 miles from the section to the La Plata.   To spread the inclination gained by the rise of
the river uniformly over so great a distance is certainly incorrect ; the fall will adjust itself
to the various reaches of the river, and the additional fall required in some localities to pass
the increased volume may be large in some and small in other reaches ; we can, accordingly,
not accurately determine the flood volume of the Paraná at Rosario without at least one surface current observation during the period of flood and the corresponding depth of water for
that observation.   We should then obtain the bulk of the flood, which would nevertheless
fall short of as much as might find its way over the flooded islands and the Victoria or
Paranacito channel at Punta Gorda, at the commencement of the delta.   The flood volume
of the Paraná might be gauged near Corrientes with considerable accuracy ; in the lower
reaches it would be difficult if not impracticable to determine its precise figure.   At exceptional rises of the Paraná, such as in 1858 and 1868, when the flood level reaches about 24 feet
above the ordinary low water of river, the whole delta is one sea, and appears an extension
of the La Plata with innumerable submerged trees and woods, some above water level
marking the position of islands.   Great as the volume of the Paraná at its lowest summer
level is—immense in comparison to the largest European rivers,—and much larger than that
of all the rivers of Europe put together—it is but a small fraction of its flood volume during
exceptional rises ; and, we can only wonder at the magnitude of the sources which for
months, nay for whole years together, pour forth inconceivable masses of sweet water, every
drop of which had been raised by the power of the sun from the Atlantic and Pacific oceans
above the tops of the highest mountains of Brazil and the Andes.

# CHAPTER VI.

## The Uruguay.

### FROM RIVER TO RIVER.

ON the 26th January, 1871, 8.30 A.M., we left the Rosario section by the 'Aguila,' and sailed for the river Uruguay. On our way we intended to gauge the volume of the Paranacito channel which escaped the Rosario section. For this purpose, instead of following down the channel of the main river, we turned into the Pavon, branching from the Paraná 17½ miles below the section. At 10.30 A.M. we entered the Pavon, but our progress was stopped by a blockading party of mariners from the National Government. This channel led to various small ports of the province of Entre-Rios, then in open rebellion against the legitimate Government; and the communication of those ports with the Paraná was cut off by the authorities. The captain of the blockading party stated he could not allow any vessel to pass without a special permit, and we had in vain explained that there were two special pilots from the Government on board, one from Buenos Ayres, the other from the Captain of the Port of Rosario, that they might search our ship, which carried neither goods nor contraband of any kind, &c. The captain, however, pleading instructions, we were in a dilemma either to give up the gauging of the Pavon, Paranacito, and Ibicuy channels, or to run the blockade. Steam having been raised above the average pressure, we sent a final message to the captain asking him to pay us a visit and ascertain for himself that we were friends and not enemies of the Government, and that " we should not be stopped." Fortunately he came, and our assistants talked matters over with him; and, having felt by degrees at home, he complained of the hardships and privations they had to undergo in blockading the Pavon channel in the midst of a wilderness, receiving but indifferent and irregular supplies of provisions; and, sometimes they had nothing else but what they could find by their guns. We thereupon offered a number of sheep and other like matters which we had in abundance, and were soon after permitted to enter the Pavon to continue our exploration, after a delay of only one hour and a half.

Although it was believed that the volume of the " Paranacito or Victoria" channel was the only one that escaped measurement at the Rosario section, we ascertained the volume of the Pavon included in the above measurement, in case other branches should join it later on. At 2 P.M. a suitable locality presented itself for the gauging of the Pavon, and our assistants cleared the ground by the aid of the sailors for a suitable base, &c., preparatory to observations. The feature of the surrounding islands of the delta of the Paraná was here the same as at other localities; from the margin of the river firm land soon declined into a marsh, and farther on into an inland lake, teeming with all kinds of waterfowl; it was impracticable to proceed in many places a hundred yards from shore. The heat was very oppressive, about 105° Fahrenheit in the shade. At 3 P.M. a storm seemed to approach

s 2

from the distance; a range of dark clouds being just visible above the horizon. At 4 P.M. the approaching storm appeared formidable; heavy clouds to the S.W. darkening the horizon, although the air was a perfect calm. At 4.30 P.M. the clouds assumed dimensions and outlines and colours which we never before saw, and we were all somewhat alarmed; the 'Aguila' was made as safe as the sailors could by additional hawsers; for the next thirty minutes we were watching, with wonder and surprise, these strange clouds, approaching with extraordinary velocity; at least at a rate of a mile a minute. Heavy storms are of frequent occurrence in South America, and on that account they are commonplace, and are looked upon with indifference. The approaching storm was, however, essentially different in appearance; for, the clouds were not a mile or two above the ground, they were touching it. Their outline was hard and defined, like those of cumuli, yet not rounded; a number of straight projections, elevations, and depressions and ridges of a dark brown and grey colour, reaching from the earth to the towering height of quite three miles. Indeed, to all appearance the Andes were rushing upon us at express speed, threatening to bury the insignificant 'Aguila' and crew among the boulders of their outskirts. At 5 P.M. the "boulders," floating about 50 feet above the ground, just burying the tops of small trees, came up with us; and in an instant we were from the light of day in the midst of night and darkness, and nothing more could be seen; the roaring of the storm alone could be heard. Total darkness lasted from five to six minutes, within which the air grew thick and choking, covering the 'Aguila' and the whole land with a deposit of fine impalpable clay to the thickness of perhaps a tenth of an inch; so much, at any rate, the storm left behind on deck after it had somewhat subsided, and the ghost of the Andes had past, and an ordinary storm of rain, with thunder and lightning, had displaced it. The temperature fell in a few minutes from 105° to 65° Fahrenheit. A deluge of rain then lasted about one hour; after which it moderated, and at 9 P.M. it had entirely passed. It was a heavy "dust storm," frequent on the Pampas during the summer; and although we had seen many, none bore likeness to this one. The intense heat on these plains lasting for weeks, nay, for months, without a drop of rain, parches the ground, which assumes the brown colour of the land and produces at the top a layer of impalpable fine clay. A gentle breeze may raise clouds of this dust to great heights; and it is probably due to this circumstance that dust storms on land do not present the grand and hard outlines of mountains, because they disturb the dust of the land in advance and make everything appear hazy; obliterating their outline. As the storm, however, passed from the table-land over the delta of the Paraná, there was no more any parched land within the basin of the old Plata estuary; and the air remaining clear, the masses of clay could assume hard outlines as they were pressed forward by the storm in their rear. And so it happened that we witnessed a sight of rare occurrence; only once to be seen, and never to be forgotten.

On the 27th January, in the morning, the section of the Pavon was made and currents observed, all of which had been completed by 11 A.M.; and we mention the general dimensions and quantities to show, that the Pavon channel, although a small matter compared to the Paraná, would rank among great rivers in Europe. The width of the river from margin to margin with steep banks was 1029 feet 6 inches at the section; the depths being uniform from shore to shore, varying over the larger part of section from 32 to 38 feet, with a mean depth of 29 feet and a sectional area of close upon 30,000 square feet. The mean surface current from bank to bank was 121 feet 3 inches, and the maximum current 168 feet

6 inches per minute. These figures determine the inclination of the Pavon to be the same as that of the Paraná, found to be 0·37 inch per mile.

After the completion of the section at 11 A.M. we continued our journey down the river, which at 2.30 P.M. was abruptly put a stop to by having run at full speed upon a bank in the middle of the river, and the keel of the 'Aguila' buried itself about 15 inches in the muddy bottom. This accident appeared a greater difficulty than the blockading squadron at the mouth of the Pavon, and neither anchors nor engine power seemed to move the boat the least. The Paraná was fortunately rising about an inch a day; there was hope that at the worst we might float again within a week or a fortnight; and we had cause to regret our generosity in parting with some of our provisions to the blockaders. The prospect of short provisions had its effect on the crew; and, with great energy and exertion everything on board ship had been landed on shore by the aid of our large boats; the whole supply of coal, iron, timber, &c.—everything that could be removed was landed to lighten the 'Aguila,' and at 8 P.M., by anchors and engine power, we got her, after much trouble, afloat again; and we remained in deep water over the night. At sunrise next morning the coal, &c., was taken on board, and at 8.45 A.M., January 28th, we again sailed down the Pavon. At 10.25 A.M. we passed the Paranacito or Victoria channel, but the locality being unfavourable for observations, the journey was continued after the junction of the two rivers, which then take the name of "Ibicuy," until at 2.15 P.M., when a suitable locality presented itself to gauge the river.

The distance from the Paraná to the Paranacito channel, which escaped the gaugings at the Rosario section, is accordingly by the Pavon 7 hours and 25 minutes' sail with the 'Aguila,' or about 52 miles along the sinuous course of the river; in a straight line it may be from 30 to 35 miles, representing in this locality the width of the delta.

At 2.15 P.M., January 28th, the Ibicuy presenting a favourable reach to gauge its volume, the usual preparatory arrangement for observations had, without delay, been made. The Ibicuy conveys the volume of the Pavon and the Paranacito rivers; the former had already been gauged the day before. The channel of the Ibicuy was on a gentle curve at the section; a straight reach could not be found. The width of river 949 feet at section; and therefore by 80 feet narrower than the Pavon; the depth of the Ibicuy was found to be, however, much greater. The mean depth was 53 feet 5 inches, and for the larger part of the section it varied from 50 to 90 feet. The sectional area came close upon 50,600 square feet. The mean surface current 100 feet 6 inches, the maximum 118 feet 9 inches, per minute; the inclination of its surface being accordingly very small, considering its gentle surface current and its great depth; at the locality of section it was about one-fourth of that of the Paraná, as determined by the aid of Diagram No. 1, Plate V. Although the surface current of the Ibicuy was smaller than that of the Pavon, its mean current nevertheless exceeded that of the latter, owing to depth alone. In this reach of the delta no tidal rise and fall of the river's surface could be traced in twenty-four hours.

At noon, January 29th, the voyage down the Ibicuy was continued, and at 5.20 P.M. we came up with another blockading party near the junction of the Ibicuy and the Paraná Guazú. The 'Dolercitas,' a powerful paddle-steamer, with cannon and about fifty mariners, were blockading the Ibicuy by order of the National Government. We were, however, well acquainted with the captain and crew of the 'Dolercitas,' having spent some days together during the survey of the Guazú; and we were looking forward with satisfaction to meet old friends. Yet, we were here in great danger. We thought the 'Dolercitas' would recognize us from a mile or so; they did not, and took our steamer for a rebel man-of-war belonging to Lopez-Jordan, then the chief of the province of Entre-Rios, and at war with the legal Government. We watched the 'Dolercitas' through our glasses on deck, and noticed a great commotion on board ship. About a mile distant, she fired her cannon across the river. We took it for a salute, as at previous meetings; and, taking no further notice, we bore straight down upon her. We could distinctly see the shifting of her cannon on deck, though we did not understand it; we were "sighted" with canister and grape, and every man on board had twenty-four rounds of ammunition, and was ready and prepared to fight. Some refugees on board the 'Dolercitas' urged the captain to fire, and not to wait until we were right on him; but, like a sailor, he declined until he could see some hostile move. At about 300 yards the captain recognized the 'Aguila,' and it would be as difficult to describe his relief and pleasure to meet friends in a hostile wilderness, as it would be to express our surprise at the position and the danger we had been in during the preceding ten minutes. We had to remain over night, and sailed again next morning at 7 A.M., January 30th, 1871. At 2.45 P.M. we arrived at Higueritas, at the mouth of the Uruguay, and having obtained provisions and a new pilot for the navigation of the river, we commenced our voyage at 6 P.M. for the exploration and gauging of the Uruguay.

### THE RIVER.

The little town of Higueritas, also called Nueva Palmira, is situated in lat. 33° 52' S., long. 58° 23' W., in the Banda Oriental, at the junction of the Uruguay with various branches of the Paraná, all of which discharge jointly their volume into the La Plata. Three miles below Higueritas, at Punta Gorda, the La Plata proper commences; three miles above Higueritas the Uruguay opens into a lake, from 4 to 6 miles wide and about 56 miles long. There are no islands on this lake, although, with the exception of a deep channel half a mile wide of steep sides and submerged, the lake is shallow; it may be called the estuary of the Uruguay. A little above Fray Bentos, 58 miles from Higueritas, the first islands appear within the lake; and, their number soon increasing, we enter the delta of the Uruguay, which for 25 miles more retains the width of the lower lake, breaking, however, up into a great number of large and small islands, until, a little below Paysandú, the river proper commences within a confined channel. At Paysandú, a commercial town of importance, 125 miles from Higueritas, the delta of the Uruguay commences; at Fray Bentos the visible delta terminates; and from the latter place to the La Plata the future delta of the Uruguay is now in course of formation.

We sailed at 6 P.M., January 31st, from Higueritas, and soon after we left this sandy place, our attention was called to the marked difference of colour of the waters we were sailing on. The western half of the channel was turbid, like all the waters of the Paraná; the eastern portion was clear, without any tint; and the demarcation between the two waters was sharp and defined, and they did not appear to mix at all. Samples were taken by crossing the line of demarcation within 50 yards of each other; the one was water from the Paraná, the other from the Uruguay; the former derived from a number of large branches to the west of Higueritas, the latter from the lake. The Uruguay water has the reputation of purity and clearness. We shall presently see that turbid water of the Uruguay proper, issues of crystalline clearness from this lake. During the survey of the Uruguay there was a periodical rise of the river, viz. on February 3rd, 1871, and a sample of water was taken on that day at the Salto section, about 200 miles above Higueritas. The water was turbid, of deep-brown colour; and the analysis of the sample showed that it contained 1 part by weight solid matter in suspension in 9524 parts of water. There was no perceptible change in the colour of the water or in its analysis, until we reached Fray Bentos on the 5th February, 1871, and here it contained 1 part solid matter in 11,200 of water by weight in suspension. At Higueritas, on the same day, the waters of the Uruguay appeared clear, and we could only trace 1 part of solid matter held in suspension by 25,925 of water. Nothing could more forcibly illustrate the formation of deltas. The river retains matter held in suspension by its water within its ordinary channel as long as its velocity is maintained; as soon as it enters a lake or an estuary checking regular currents, the matter held in suspension is dropped. On the same day a sample was taken of the turbid water on the western portion of the channel in front of Higueritas, derived principally from the Gutierrez and Bravo branches of the Paraná; it was found to contain by weight 1 part solid matter in 4586 parts of water in suspension. Two months previously the Guazú, a little below the junction of the Ibicuy, held only 1 part in 9000 of water in suspension, showing how much the proportions may vary according to locality and the seasons of the year.

At 10.45 A.M., January 31st, we passed Fray Bentos, 58 miles above Higueritas, having during the night stopped several hours. The bluff is to all appearance similar to that of the right bank of the Paraná at the Rosario section, and of the tosca formation. On the top of the bluff there are large works for the manufacture of the celebrated " Liebig's Extract of Meat." Above Fray Bentos we enter among numerous very pretty islands, thickly wooded with a great variety of trees. The islands are about 4 feet above the level of the water, and are different in appearance from those of the delta of the Paraná. The speed of the 'Aguila' up the Uruguay was inconveniently small as against the shore, indicating a stronger current than in the Paraná. Paysandú was reached at 10.45 P.M.; our intention was to gauge the river somewhere about here if the effect of tide should not be felt. A gauge was put down at midnight; by 6 A.M. the river rose 2 inches; by 8 A.M. 3 inches more; in all a rise of 5 inches since midnight. It was also ascertained from information of fishermen that the river at Paysandú is daily rising and falling, in other words, that it is affected by the tides of the La Plata. It was also observed that the Uruguay assumes a decided tint, and that 15 miles above Paysandú it was in appearance much like the water of the Paraná, a yellow-brown, getting more and more opaque; sample of water having been taken for analysis February 1st, noon. At 3.15 P.M. we observed the cliffs on the right bank of river to be solid rock, about 100 feet high, and that palm trees were getting numerous; and we see the first time palm groves and forests on the right bank. At 5 P.M. the water of the Uruguay was browner than that of the Paraná, assuming a sepia tinge. The heat had been all day oppressive, 97° Fahr. in the shade. At 10 P.M. we passed some rapids on the river, and at 2 A.M., February 2nd, the town of Salto, about 200 miles above Higueritas, was reached; it is a port of importance on the Uruguay.

The navigation of the Uruguay practically terminates at the port of Salto, except during great rises, when the Falls above Salto may be passed by powerful steamers. There is a small Fall about a mile above the town, called "Salto Chico," which during the periodical rises is submerged, forming rapids. The current at Salto was at the time (2nd February, 1871) strong; the paddle-wheels of the 'Aguila' had to be fastened by ropes to prevent the current turning them round. We then sailed in search for a suitable locality to gauge the Uruguay, which we found about a mile and a half above "Salto Chico," then submerged; after a series of preliminary soundings, varying from 18 to 30 feet depth, and disclosing a uniform bed within a straight reach of the river for several miles—the locality for the section had been determined, and operations had at once commenced.

The Uruguay is a great river, subject to extraordinary fluctuations. During our surveys in the beginning of February, 1871, the river was in flood, having been in one of its periodical rises, which commenced about the middle of January. It was above the average January rise by 6 feet, being at the town of Salto 22 feet on ordinary low water. The January rise usually lasts about three weeks. The next periodical rise of the river occurs in April, which exceeds the ordinary January rise by 12 to 14 feet, measuring 30 feet on ordinary low water at the town of Salto; this second rise generally lasts two months. The great flood of the Uruguay always happens in September and October, and high water lasts a whole month, and rises from 45 to 50 feet on ordinary low-water level at Salto. The rise of the river is slow at first, but rapidly increases, and often amounts to 3 feet in one day; the fall is comparatively slow. The flood of September and October is certain and regular; the periodical rises are not so regular either in height or time of their occurrence. The year 1869 was exceptional, the flood level being maintained for ten months, from January to the end of October. The river is always very low in December. There are during the year several minor changes; all great rises are, however, independent of the rainfall on Argentine territory or on that of Uruguay; the floods derive their whole volume from the mountains of Brazil. The drainage area of the Uruguay is under 200,000 square miles; it is not even approximately known. During low water the volume of the Uruguay is comparatively insignificant. We ascertained from a reliable source that during December, when the river is very low, the whole volume of the Uruguay is confined within a rocky channel about two miles below Salto, called the "Corralito"; the width of the channel is stated to be 145 feet, with a depth of water of 6 feet, and a current of about five miles an hour; the length of channel about 700 feet. These data would reduce the dry weather volume of the Uruguay to a small percentage of its flood volume of October. Although we had no opportunity of seeing the Corralito, as it was submerged at the time by the periodical rise, it will be seen by reference to the Salto section, Plates V. and VII., that the low-water area is comparatively small, and the current must be very feeble; because a bar of rocks, about $1\frac{1}{2}$ mile below the section, called Salto Chico, runs across the bed of the river, stopping all navigation, and consequently the indication of the river's surface can only be very little. During the surveys these rocks were submerged, and steamers of ordinary size might have passed over them anywhere.

About nine miles above the town of Salto there are the great Falls of the Uruguay, called "Salto Grande." A succession of reefs cross its channel diagonally, making it at the Falls about $1\frac{1}{2}$ mile wide. When the river is low there is a clear drop of from 12 to 15 feet, and the difference of level above and below the Falls is about 25 feet. During the floods the rocky bar is submerged, forming a succession of rapids about 1½ mile long, similar in appearance to those of the Niagara. On the right bank there is a rocky channel, which during the October floods may be ascended by powerful steamers; although, at the time of our visit, this channel was blocked by the wreck of a steamer of considerable size; illustrating the difficulties and the danger to navigation. Except during the September and October flood, for about six weeks during the year, Salto Grande closes all navigation between the upper and lower reaches of the river. Above the Falls the navigation is easy during the periodical rises, and would be of importance to the trade with Brazilian ports. Above Salto Grande the rise of the Uruguay due to the October flood amounts to 12 feet on its low water; this rise above the Falls produces one of from 45 to 50 feet at the town of Salto, only about nine miles below the Great Falls.

The "Salto Chico" Fall, about a mile above the town, and consequently 8 below the Great Falls, closes all navigation during six months of the year; but during the periodical rises and the October floods of the river, it may be easily passed by steamers drawing from 4½ to 5 feet, and its navigation is practicable during six months of the year. The removal of the rocks, if it were desirable for other considerations affecting the levels of the river, would be difficult and expensive; they are very hard, and belong to primary formations.

Vessels drawing 4½ feet may at all times ascend the Uruguay as far as the town of Salto, about 200 miles above the junction with the La Plata; during the periodical rises or the October floods the river is deep. Whenever the volume of the Uruguay is considerable, due to small or great floods, its water in the upper reaches is always turbid and of sepia colour, holding a considerable amount of solid matter in suspension. During our journey down river on the 5th February, 1871, there was no change perceptible in the colour of the Uruguay from Salto to near Paysandú; it appeared a dirty brown; it then assumed various shades resembling by degrees the water of the Paraná; and, at Fray Bentos, it was getting transparent, at Higueritas of crystalline clearness. For the present the Uruguay contributes next to nothing to the formation of banks and islands of the La Plata, which will be entirely due to deposit from the Paraná. The Uruguay has a large basin of its own to deposit matter which it holds in suspension as far as Paysandú. At the latter place its delta commences, of which about one-half, as far as Fray Bentos, had been reclaimed from the lake; the other half, as far as Higueritas, is now its basin for settlement and deposit.

The Uruguay is at times a mighty river, rivalling the Paraná; at others it sinks into comparative insignificance. The Paraná is a great river at all times, conveying an immense volume to the Atlantic at its lowest condition, never falling to 50 per cent. of its ordinary flood volume. The Uruguay is subject to frequent and remarkable fluctuations; during the great floods it rivals the Paraná, and one month later it sinks to a small percentage of its former greatness. The Paraná is the type of a truly great river; the Uruguay represents a mighty torrent of extraordinary dimensions.

T

## SALTO SECTION.

At the locality of section the Uruguay is confined between high banks, and the river's channel is straight for several miles. The left bank consists of a continuation of rocky bluffs, 60 feet above ordinary low water, descending almost in a vertical line at the margin of the river. The right bank slopes into the water at an inclination of about 1 vertical in 6 or 7 horizontal. The line of base was measured on the left shore on the top of the rocky bluff, the right bank being thickly covered with woods to the margin of the water. Although the currents were very strong, there was not a ripple or a whirlpool to be seen, and from the top of the bluff the outline of every cloud could be seen reflected from the river's surface; all of which indicate an even bottom and no disturbance of currents. There was calm weather with intense heat. Some difficulty was experienced with the selection of the base, as it could not be obtained with sufficient length close to the margin of the bluff on account of the undulating surface of the ground; and, if placed farther inland, the flags appeared covered by the bluff and the trees for near points. Two lines were accordingly adopted; one at the margin of the bluff, 139 feet 5 inches, for near points; the other, the main base of 600 feet, for greater distances from shore. There were groves of palm trees on the right bank, and the river viewed from the base presented a very fine appearance, differing materially from that of the Paraná in similar latitudes.

### DIARY OF OBSERVATIONS.

#### FEBRUARY 2ND, 1871.  LOCALITY: SALTO SECTION.

Lat. 31° 21' S., Long. 57° 55' W.; 1½ mile above Salto Chico ; 2½ miles by river above the town of Salto.

*Soundings from board the 'Aguila.'*

| No. of Sounding. | Angle on Line of Base. | Sounding in Feet and Inches. | Remarks. |
|---|---|---|---|
| | ° ' | ft. in. | |
| No.  1 | 58 43 | 25  3 | 4 P.M., February 2nd, 1871.  Gauge, 4 feet 6 inches. Angles on base of 600 feet. |
| „    2 | 45 44 | 30 11 | |
| „    3 | 36  4 | 29  5 | |
| „    4 | 28 45 | 34  6 | |
| „    5 | 23 27 | 28  8 | |
| „    6 | 20 30 | 29  9 | |
| „    7 | 18 18 | 28  4 | |
| „    8 | 15 33 | 26  7 | |
| „    9 | 14 36 | 24  5 | |
| „   10 | 13 53 | 21  2 | |
| „   11 | 12 46 | 18  3 | |
| „   12 | 12 51 | 13  1 | |
| „   13 | 56 51 | 24  4 | 5 P.M.  Sounding completed.  Gauge, 4 feet 6 inches. |

FEBRUARY 4TH, 7 A.M.  *Additional Soundings.*

River had fallen 12 inches since No. 13.

| | | | |
|---|---|---|---|
| „   14 | 28 43 | 25  4 | Gauge, 3 feet 6 inches.  7 A.M., February 4th.  Angles |
| „   15 | 22 24 | 24  1 | on auxiliary base, 139 feet 5 inches long. |
| „   16 | 14 11 | 29 10 | Completed, 7.15 A.M. |

MEMORANDA.

**Base** on left bank of river, 600 feet by steel tape. Auxiliary base, 139 feet 5 inches at margin of bluff.

Gauge: At noon, February 2nd, 1871, 4 feet 6 inches; at 6 P.M., 4 feet 7 inches. February 3rd, 6 A.M., **4 feet 6** inches; 1 P.M., 4 feet 4 inches; 6 P.M., 4 feet 2 inches. February 4th, 6 A.M., 3 feet 6 inches.

Flood level: The October flood level was pointed out on rocks by a native of Salto at two places separated by a hill;—found by levelling, on February 3rd, at noon, that the level of the two indicated points was the same; they were some distance inland, and 20 feet 2 inches above the present level of the Uruguay. Base at A, 39 feet 6 inches above level of river, February 3rd, noon.

Section: At right angles with base and at right angles with direction of current.

FEBRUARY 3RD, 1871. LOCALITY : SALTO **SECTION. WEATHER PERFECT CALM. CURRENT OBSERVATIONS, METER No. 1.**

No. a.  Angle, 28° 44′, on auxiliary base 139 feet 5 inches.

Time $\begin{cases} 2\text{h. }30\text{m. }0\text{s. commencement} \\ 2\text{h. }35\text{m. }0\text{s. termination} \end{cases}$ 5m. 0s. observation.  Indexes $\begin{cases} 0a + & 0 \\ 8a + & 20 \end{cases}$ 1628 revolutions.

Check observation .. .. .. .. 1m. 0s.  „   „ $\begin{cases} 8a + & 20 \\ 9a + & 144 \end{cases}$ 335  „

Angle after observations, 28° 22′; take mean, 28° 33′.

**No. b.**  Angle on base A B, 600 feet, 47° 11′.

Time $\begin{cases} 2\text{h. }58\text{m. }0\text{s. commencement} \\ 3\text{h. }3\text{m. }0\text{s. termination} \end{cases}$ 5m. 0s. observation.  Indexes $\begin{cases} 0a + & 0 \\ 7a + & 150 \end{cases}$ 1557 revolutions.

Check observation .. .. .. .. 1m. 0s.  „   „ $\begin{cases} 7a + & 150 \\ 9a + & 54 \end{cases}$ 306  „

Angle after observations, 47° 11′.

**No. c.**  Angle on A B, 35° 30′.

Time $\begin{cases} 3\text{h. }31\text{m. }0\text{s. commencement} \\ 3\text{h. }36\text{m. }0\text{s. termination} \end{cases}$ 5m. 0s. observation.  Indexes $\begin{cases} 0a + & 0 \\ 7a + & 140 \end{cases}$ 1547 revolutions.

Check observation .. .. .. .. 1m. 0s.  „   „ $\begin{cases} 7a + & 140 \\ 9a + & 50 \end{cases}$ 314  „

Angle after observations, 34° 35′.

NOTE.—Boat oscillates to and fro 1°. Angles increasing and again decreasing regularly; take mean 35°.

**No. d.**  Angle on A B, 27° 25′.

Time $\begin{cases} 3\text{h. }59\text{m. }0\text{s. commencement} \\ 4\text{h. }4\text{m. }0\text{s. termination} \end{cases}$ 5m. 0s. observation.  Indexes $\begin{cases} 0a + & 0 \\ 8a + & 0 \end{cases}$ 1608 revolutions.

Check observation .. .. .. .. 1m. 0s.  „   „ $\begin{cases} 8a + & 0 \\ 9a + & 108 \end{cases}$ 309  „

Angle after observations, 27° 25′.

No. e.  Angle on A B, 20° 32′.

Time $\begin{cases} 4\text{h. }26\text{m. }0\text{s. commencement} \\ 4\text{h. }31\text{m. }0\text{s. termination} \end{cases}$ 5m. 0s. observation.  Indexes $\begin{cases} (0a + & 0) \\ (7a + & 60) \end{cases}$ 1467 revolutions.

Check observation .. .. .. .. 1m. 0s.  „   „ $\begin{cases} (7a + & 60) \\ (8a + & 153) \end{cases}$ 294  „

Angle after observations, 20° 28′; mean, 20° 30′.

**No. f.**  Angle on A B, 16° 38′.

Time $\begin{cases} 4\text{h. }50\text{m. }0\text{s. commencement} \\ 4\text{h. }55\text{m. }0\text{s. termination} \end{cases}$ 5m. 0s. observation.  Indexes $\begin{cases} (0a + & 0) \\ (6a + & 163) \end{cases}$ 1369 revolutions.

Check observation .. .. .. .. 1m. 0s.  „   „ $\begin{cases} (6a + & 163) \\ (8a + & 31) \end{cases}$ 270  „

Angle after observations, 16° 28′; mean, 16° 33′.

**No. g.** Angle on A B, 14° 20'.

Time $\begin{cases} \text{5h. 13m. 0s. commencement} \\ \text{5h. 18m. 0s. termination} \end{cases}$ 5m. 0s. observation.   Indexes $\begin{cases} (0a + & 0) \\ (6a + & 62) \end{cases}$ 1268 revolutions.

Check observation .. .. .. .. 1m. 0s.   „     „   $\begin{cases} (6a + & 62) \\ (7a + & 114) \end{cases}$ 253   „

Angle after observations, 14° 14'; mean, 14° 17'.

---

**No. h.** Angle on A B, 13° 1'

Time $\begin{cases} \text{5h. 36m. 0s. commencement} \\ \text{5h. 41m. 0s. termination} \end{cases}$ 5m. 0s. observation.   Indexes $\begin{cases} (0a + & 0) \\ (5a + & 128) \end{cases}$ 1133 revolutions.

Check observation .. .. .. .. 1m. 0s.   „     „   $\begin{cases} (5a + & 128) \\ (6a + & 155) \end{cases}$ 228   „

Angle after observations, 13° 0'; mean, 13° 0½'.

---

**No. k.** Angle on A B, 12° 28'.

Time $\begin{cases} \text{5h. 55m. 0s. commencement} \\ \text{6h. 0m. 0s. termination} \end{cases}$ 5m. 0s. observation.   Indexes $\begin{cases} (0a + & 0) \\ (2a + & 50) \end{cases}$ 452 revolutions.

Check observation .. .. .. .. 1m. 0s.   „     „   $\begin{cases} (2a + & 50) \\ (2a + & 145) \end{cases}$ 95   „

Angle after observations, 12° 28'.

This position is over submerged willows.

---

### MEMORANDA.

Angle from margin of water of right bank on 12° 5' on A B; there is a current to the very margin of river, although for about 50 feet there are many willows submerged.

Margin of left bank 161 feet 9 inches from base, with a strong current. Weather a perfect calm, surface of river like a mirror. The currents across the river on line of section are very strong, and the mooring anchors frequently drag for some time until they finally grapple and the boat remains on line of section oscillating to and fro as indicated by the sextant.

By the aid of the auxiliary base, soundings close to the left shore were repeatedly taken; not, however, without difficulty, on account of the current carrying the 'Aguila' rapidly from the intended point; the steamer appeared helpless for prompt movements. A rowing boat could not ascend the river.

Sailed from the Salto Section, 8.15 A.M., February 4th, 1871, to the town of Salto; the port being reached in sixteen minutes; at 11.24 A.M. again ascended the Uruguay as far as the Great Falls called "Salto Grande," which were reached 1.50 P.M. The 'Aguila' could not ascend the channel of the rapids; it was blocked by the wreck of a large steamer.

The above observations may be worked out according to the rules which have been followed in the preceding chapters, and which have been fully explained. In the current observations we were guided only by the five-minute trials to determine the velocity of current in feet per minute by the equation of the meter; provided the minute observations, as a check on the first, confirmed the general accuracy of the former. The result of the observations of the Uruguay at the Salto section will appear from the following Tables :—

TABLE IX.

FEBRUARY 2ND, 1871. THE URUGUAY; SALTO SECTION.

*Soundings on Line of Section* (from board the 'Aguila').

| No. of Sounding. | Depth in Feet and Inches. | Distance of Sounding from A of Line of Base, in Feet and Inches. | Time. | Gauge. |
|---|---|---|---|---|
|  | ft. in. | ft. in. |  | ft. in. |
| No. 1 | 25  3 | 864  7 | { 4 P.M. { February 2nd, 1871 } | 4  6 |
| „  2 | 30 11 | 584 10 |  |  |

TABLE IX.—(continued).

| No. of Sounding. | Depth in Feet and Inches. | Distance of Sounding from A of Line of Base, in Feet and Inches. | Time. | Gauge. |
|---|---|---|---|---|
| | ft. in. | ft. in. | | ft. in. |
| No. 3 | 29 3 | 824 4 | | |
| „ 4 | 34 6 | 1093 8 | | |
| „ 5 | 28 8 | 1383 2 | | |
| „ 6 | 29 9 | 1604 9 | | |
| „ 7 | 28 4 | 1814 3 | | |
| „ 8 | 26 7 | 2156 3 | | |
| „ 9 | 24 5 | 2303 5 | | |
| „ 10 | 21 2 | 2489 7 | | |
| „ 11 | 18 3 | 2648 0 | | |
| „ 12 | 13 1 | 2702 8 | | |
| „ 13 | 24 4 | 391 10 | 5 P.M. February 2nd, 1871 | 4 6 |
| „ 14 | 26 4 | 324 4 | 7 A.M. February 4th, 1871 | 3 6 |
| „ 15 | 25 1 | 408 1 | | |
| „ 16 | 30 10 | 621 6 | 7.15 A.M. February 4th, 1871 | 3 6 |

MEMORANDA.

Margin of right bank from A of base, 2802 feet 9 inches.

„ left „ „ 161 „ 9 „

Width of river on line of section, margin to margin, 2641 feet.

One foot was added to Nos. 14, 15, and 16, to reduce soundings to the same level.

TABLE X.

FEBRUARY 3RD, 1871. SALTO SECTION. SURFACE CURRENTS.

Currents across River on Line of Section.

| No. of Trial. | Time of Observation. | Current in Feet per Minute. | Distance of Locality of Observation from A of Line of Base. |
|---|---|---|---|
| | h. m. s. | ft. | ft. in. |
| No. a | 2 22 30 P.M. | 335·1 | 326 1 |
| „ b | 3 0 30 „ | 319·4 | 555 9 |
| „ c | 3 38 30 „ | 317·4 | 856 10 |
| „ d | 4 1 30 „ | 329·8 | 1156 8 |
| „ e | 4 28 30 „ | 301·4 | 1604 9 |
| „ f | 4 52 30 „ | 281·7 | 2019 1 |
| „ g | 5 15 30 „ | 261·4 | 2356 9 |
| „ h | 5 38 30 „ | 234·4 | 2597 2 |
| „ k | 5 57 30 „ | 97·6 | 2713 11 |

MEMORANDA.

Position k over submerged willows; will somewhat reduce velocity. Current considerable at the margin of river; on the left bank very strong and agitated from rocks.

## ANALYSIS OF SECTION.

The section of the Uruguay is represented on Plate VII., horizontal and vertical scale being the same, viz. 1 inch of drawing to 100 feet actual dimensions. From the base A, at a distance of 110 feet, the table-land of the Republic of Uruguay abruptly falls 40 feet to the periodical flood-level of the river; at a distance of 161 feet 9 inches from A its effective margin is reached, from which point, by another abrupt descent, the bottom of the river is found about 30 feet below the water level. After the auxiliary base had been adopted, an attempt was made to get soundings as near as possible to the margin, but none could be obtained nearer than No. 14; for, it was necessary to sight two flags on the line for direction; this circumstance accounts for the numerous soundings which crowd the locality near the left bank. The depth of the Uruguay is remarkably uniform for so wide a river, and for 2200 feet it oscillates between 24 and 34 feet. Within those limits its bed rises and falls by gentle gradients from the left bank to about midway; its outline is accordingly not regular. From the middle to the right bank for about 1000 feet the bottom is regular, rising without intermission at gentle gradients of about 1 in 100; until within 200 feet from shore the rise is more and more rapid. On the 2nd February, 1871, noon, the sectional area of the Uruguay at the Salto section was 71,200 square feet, representing one of the river's periodical rises. At the regular October flood level the sectional area at the locality would be 126,800 square feet; and at ordinary low-water level it may be taken at 25,000 square feet. The mean depth of the Uruguay over its effective width of 2641 feet was 26 feet 11 inches on the 2nd February; during its regular October flood level it would be 44 feet 11 inches over a width of 2823 feet. The centre of gravity of section is 1370 feet 0 inch from A, and it is 14 feet 1 inch below the surface. The channel of the Uruguay at the Salto section is straight for more than two miles above, and one mile below, section; this had been ascertained by triangulation of a number of points on the right bank from the base. The Uruguay is confined between the high banks of the table-land of Entre-Rios and the Banda Oriental, which it never floods. Its shore is rocky on both sides, and of primary formation, the tertiary Pampas deposit being in contact with the primary rocks, without the intervention of the tertiary strata which underlie the Pampas deposit in the basin of the La Plata and the Paraná. The Uruguay cannot materially change its present channel from the great Falls of Salto as far as Paysandú, where the river's delta commences; within the delta changes must necessarily take place, and they do take place, and are comparatively rapid.

## ANALYSIS OF CURRENTS.

During the periodical rises, or during the regular floods of September and October, the currents of the Uruguay are very strong. Referring to the outline of surface currents of the Salto section on Plate V., it will be seen that it bears no resemblance to that of the Paraná, shown on the same sheet, except that they are both irregular and defined by a sinuous curve. During the analysis of the Paraná currents of the Rosario section, we have frequently alluded to those of the Uruguay, and the analysis of the former substantially includes those of the latter. To avoid repetition, we refer to page 117 for matters of detail and for explanation of the various terms and their meaning, which may appear in the present summary on the Uruguay currents.

If we plot the currents at their corresponding depths on the principle and in the manner adopted for the Paraná, page 116, Diagram No. 2 arises, shown on Plate V. From 27 to 33 feet depth there are a crowd of velocity-points close together, which determine the direction of the velocity-line O B of the Uruguay. All velocity-points appear to vibrate close to the right and left of that line, the exception being a, for which the depth appears too small for the corresponding velocity. Referring, however, to the section of Plate VII.

and also Plate V., it appears from the numerous soundings close to a, that there is at that place a local elevation in the bed of the river, and it is probable that the elevation under a is not maintained for any considerable distance in a longitudinal direction. We already know, that surface currents are guided or controlled by depths in a longitudinal direction preceding the section, and not by local depressions or elevations. In the Paraná we had also one point h which deviated considerably from its velocity-line, see Diagram No. 1, Plate V.; there it was caused by a local depression, here it is produced by a local elevation. The more irregular and uneven the river's bed may be, the wider must necessarily appear the variation of velocity-points from the velocity-line of river; this is in the nature of the subject, and forms the basis of the rule which had been derived; it does not call it in question as long as the velocity-points remain equally divided by a straight line. A parabola had also been passed through point c, which appeared the centre of movement, that is, the depth happened there to be the mean for a considerable distance above the section. It will be seen that, although a parabola might from 20 to 33 feet represent the law of movement without serious deviation, it nevertheless does not fairly divide the points, one group appearing above the other below the branch. Omitting, however, points a and k, a parabola would come near the experimental velocities from 20 to 33 feet depth, and it would be doubtful which should represent the movement. The fact is, that at those velocities and depths the branch of the parabola comes near a straight line, and one appears as good as the other. The point k, however, which falls half-way these depths, would settle the question against the parabola, if one point could be permitted to do so. It is for this reason that the Uruguay, although a great river of uniform section, could not have finally determined the law between "depths and surface currents," although a consideration of the centres of gravity of section and surface currents should, even with the sceptical, have left little doubt on the subject; since, if at a given inclination currents were governed and would increase or decrease as the square root of depths, or of the mean depths, the centres of gravity of section and of currents could, in a straight channel, never have fallen in the same line; because, the position of the centre of gravity of section depends, at equal distances, on the depth; and, if with the currents it would depend on the square root of depth, the centre of gravity of currents must inevitably fall towards the shallower portion of the section; it would be the farther from it the greater the difference in the depths may be. The Rosario section of the Paraná, however, settled the question. The position of centres of gravity and disturbance arising from bends, traced by the same law, &c., are only confirmatory evidence of its truth.

The area described by the surface currents at the Salto section on the 3rd February, 1871, was 755,000 square feet in one minute; nearly the same as that of the Paraná at low water, being only 11,000 square feet short. The mean surface current of the Uruguay from margin to margin over a distance of 2641 feet was 285 feet 8 inches per minute. If we plot the latter current on Diagram No. 2 to the mean depth of river equal to 26 feet 11 inches, the "Uruguay mean" falls right on the velocity line previously determined; and in the analysis of currents of the Paraná it has been shown that this does not follow as a matter of course. On the 3rd February the surface currents reproduced their corresponding depths in 5⁴⁄₇ seconds of time at the section, whilst in the Paraná at the Rosario section the currents required 13⅓ seconds to reproduce their depths. The Uruguay is therefore a much more rapid river; its surface velocity being 2½ times that of the Paraná, and this comparison is absolute, taking the Rosario and Salto sections to represent the two rivers. This notation might be usefully adopted for all rivers; it is short and precise. We may have a river of $\frac{1}{2}$, 1, 2, 3, . . . seconds velocity, meaning that in the space of time named the surface current travels a distance equal to the depth at the locality; it measures the absolute and relative velocity of the rivers. It is independent of particular depths or dimensions, and is the only direct measure of the effective inclinations of the rivers' surfaces at the localities under consideration.

The maximum surface current at the Salto section is close to the left bank, and was found to be 333 feet 1 inch per minute; the current then decreases, and within about one-fourth the river's width it attains a local minimum of 315 feet per minute, again increasing to nearly its maximum velocity before the middle of the river is reached, from which point it decreases at a very uniform rate for a distance of 1200 feet, when, within about 200 feet from the right bank, it falls rapidly to the margin of the river. The maximum current within the channel of a river is not by any means necessarily midway; it may be anywhere, near the right or left bank, or also midway; within a straight reach it will be always found at the greater depth which had been maintained longitudinally; and, within bends, it will be displaced from its legitimate place, and will be carried towards the concave bank of river. It has been shown during the analysis of the Paraná, that bends within a channel cause the greatest disturbance to all currents.

## ANALYSIS OF INCLINATION.

Many pages have been devoted in the preceding chapters to the delicate question of inclination of a river's surface, and we refer to pages 51; to 72; to 120; the IIIrd, IVth, and Vth chapters, respectively. It has been shown that, with a few exceptions, levelling does not disclose the effective inclination of a river's surface at a given locality, and that in nine cases out of ten it will be erroneous to assume, that the mean fall between two points along the course of a river will be also the fall at the locality. The current alone is a correct measure of the inclination of a river's surface at a given point. And, although we cannot determine from its velocity the fall without the aid of a formula, we may, from the velocity of the current referred to its depth, at once conclude, with extreme accuracy, whether the inclination in a variety of localities will be greater or smaller than at a certain point; and even determine their relative values with sufficient approximation for ordinary purposes, see pages 122 to 124.

With a river like the Uruguay, having a great many small rapids within a few hundred miles, the mean inclination between two distant points, determined usually by the convenient though inaccurate mode of barometric measurements, would be greatly misleading; between the rapids the mean fall would be considerably too great, and at the rapids too small. During great floods, submerging most of the rapids, there may be a good approximation to the effective inclination, although barometric measurements are but rude means to ascertain delicate matters. The inclination of a river's surface is a matter of importance when its improvement is under consideration, either by the removal of obstructions or shoals, their consequent effect on the upper and lower reaches, &c. &c., or by diverting a portion of its water by other channels; or, when its inundations are to be reduced or altogether to be prevented. There are many other questions referring to river improvements, to irrigation, &c., where the effective inclinations are of paramount importance. A survey cannot be considered complete without the determination of the river's gradients along its course by the aid of numerous current measurements. Having already found, by the aid of Diagram No. 2, that the velocity of the Uruguay at the Salto section on the 3rd February, 1871, was 2½ times that of Paraná at the Rosario section, we know that, inasmuch as the inclination —expressed by the difference of level between two points a certain distance apart—will approximately increase or decrease as the square of the velocities, the fall of the Uruguay at the Salto section will be about 5¼ times that of the Paraná at Rosario, or about 2 inches per mile. We must, however, bear in mind, that it is always undesirable to conclude from small quantities upon large ones, and that wherever practicable the reverse should be followed, and we should work from the larger ascertained quantities upon the smaller ones by calculation. All observations can only be a near approximation to truth, and we must always allow a margin for deviation, positive or negative. This margin of deviation may be an insignificant quantity in reference to the result of the observation; if it be, however, multiplied five or ten times, it may become a serious deviation from the truth, and vitiate the calculation. If, for example, the fall of the Paraná were accurate within $\frac{1}{12}$th inch per mile,—a very small quantity which would escape the best levels,—it would affect the fall of the Uruguay, as determined by calculation, by half an inch; and although the E wave of the La Plata determined the fall of the Paraná with extreme accuracy, see page 120, it would have been preferable to measure the fall of the Uruguay within a favourable reach and, by the unerring guide of currents, to ascertain that of the Paraná by calculation from the result of the Uruguay observations. The Salto section had been within such a favourable reach, and had we not had by arrangement to meet the February packet for Europe, we should have determined the fall by levelling. Such observations are not matter of one day; it is not merely a question of an exact levelling for several miles with a series of check levels, but also a question of the condition of the river, which should be at the time stationary. Referring to the diary of the survey, it will be seen that from the 3rd to the 4th February the Uruguay fell 12 inches, and with such oscillations of greatly increasing or decreasing quantities of water the fall could not be properly ascertained. A favourable day with a stationary level and calm weather should be selected for the purpose. During low-water level the fall of the Uruguay between the rapids will be but slight, much under one inch; during the highest floods of September and October it may be as much as three inches within the regular reaches of the river, presenting, for a considerable distance, uniform width and depth. Over the rapids themselves the fall may amount to many feet per mile, and would be matter for observation in each particular locality. At the rapids called "Corralito," for example, the dimensions of which have been already given, there is, during low water, a difference of level of 4 feet within 700 feet distance, or a fall at a rate of 30 feet per mile.

## CENTRES OF GRAVITY.

The centres of gravity of section and of currents fall nearly in the same vertical line with the Salto section of the Uruguay. The subject of centres of gravity has been fully considered at the analysis of the Rosario section of the Paraná, and the Uruguay has been referred to for comparison and argument. The horizontal distance of the centre of section from the base A B is 1370 feet 0 inch: that of the currents 1362 feet 6 inches from the same

U

line; the horizontal distance between the two being 7 feet 6 inches. Considering the great width of the river, the difference is but a small quantity, and within the margin of possible deviation. The two centres fall substantially on the same line, and although we do not wish to explain away the small difference, we may nevertheless observe that current observation k falls a little short of the velocity which would have established itself had the bottom of the river at that locality not been obstructed by submerged bushes and willows, as noted in the Survey Book during the trial of current. The agreement between the two centres determines the favourable situation of the Salto section; that the river's course had been straight for a great distance above the section; and, that nothing interfered seriously with the regular formation of currents, which had adjusted themselves on their various lines to the mean depth over which they had to pass. The centre of gravity of section is 14 feet 1 inch below the surface, and that of the surface currents 148 feet from the line of section in a horizontal plane. We observe, that centres of gravity do not fall at half the mean depth or at half the mean surface currents, but with rivers within their natural and irregular channels they will be found somewhat in advance of those mean points.

# CONCLUSION.

The survey of the Great Rivers disclosed a series of facts which we believe to be of consequence to the science of Hydraulic Engineering. The result of the surveys places the question beyond doubt, that the accepted principles upon which we were basing our arguments, when the movement of water in open or confined channels had been under consideration, were at variance with those of nature. In the analysis of the Paraná we have shown, that rivers of average dimensions do not afford the variety of circumstances which might enable us to trace the dependence between the various quantities which affect their movement; and, that even with the Great Rivers we had surveyed, there was but one locality which clearly established the law of dependence without the interference of third and fourth quantities. Dubuat evidently felt the difficulty in his masterly treatment of the subject; for, having on certain assumptions, which appeared to be warranted, framed a simple formula expressing the dependence between the various quantities, he could not make it agree with the result of observations under varying circumstances without the introduction of an arbitrary logarithmic function, which so moulded the formula that, within the limits of his trials, it appeared satisfactory. It is not, however, only a question whether or not a formula will give correct results; it is, we believe, of much greater importance that the principles which guide our thought and argument in all river improvement should be true and correct. We ought to know what happens within the channel of a river, and what is the mode of its actual movement; it is of comparatively little importance whether the fall of a river be so many inches per mile, its volume so many feet per minute; these deserve attention in their way, just as it is of importance that arithmetical operations should be correct, although they are subordinate to the functions which had determined their respective position and the values of the figures.

Our observations on the Great Rivers disclosed and established the dependence between depths and currents; that at a given inclination surface currents are governed by depths alone, and are proportional to the latter. It is impossible to exaggerate the value and the meaning of this fact. We may at once compare the inclination of essentially different localities in the same or in different rivers; and the current will be not only the most delicate, but it is the only test of the effective inclination at a given locality. If we combine with this first principle the second, disclosed by the same observations, that the current at the bottom of a river increases more rapidly than that at the surface—we have the two main points of a strong basis which will carry us over many a difficulty. If we further combine with the two preceding points the third, that for the same surface current the bottom current will be greater with the greater depth; and if we add the fourth, that the mean current is the arithmetical mean between that at the surface and bottom,—we have enough light on the subject clearly to see our way, and to comprehend what may actually happen within a river, without the aid or the supposed mysterious operation of formulæ.

The greatest current is at the surface, the smallest at the bottom; as the depth, however, increases or the surface current becomes greater, so the difference between surface and bottom currents will be smaller and smaller, approaching equality; until in great depths and in strong surface currents they are substantially alike.

The whole resistance to the movement of a river arises from its bed. Our best authors so understood it, which the construction of their original formulæ proves. It had, however, also been thought, that part of the resistance arises from friction between the liquid particles of water. It is of some importance to bear in mind, that the resistance arises entirely between a solid and a liquid; and, that to a great extent the friction between the particles of the liquid is the effect of the former, and does not cause or originate resistance to movement.

Formulæ are but the precise expression of a definite thought embodying the dependence between two or more quantities. It takes usually much training before we get used to the precise definition of thought, which also forms the language of mathematics. Like all languages mathematics is soon forgotten without practice, although it may be quickly regained by those who once possessed it. The principles derived from the survey of the Great Rivers might be expressed and defined mathematically. We have not done so for the present, because a formula cannot be considered satisfactory unless the process by which it had been derived is clearly shown; its application to numerous examples explained, and the exceptions considered; all of which, however, make up a separate treatise on the theoretical consideration of the result of the surveys; and these, we believe, should not be mixed up with the numerous facts placed on record in the preceding pages. The latter will supply enough material for the contemplation of those who love their profession for its truth and beauty; and there should be no doubt or suspicion, that facts had been moulded to suit individual views or opinions, to illustrate the accuracy of a theory which superficially considered may appear correct, though it might rest upon an imaginary basis.

In the preceding chapters we have not given any figures expressing the volume of the Great Rivers in their varying condition, because their correct expression is not practicable without a formula. It is, however, easy enough to make a close approximation if the figures should be of interest. We know from the trials on record that, for example, on the Paraná in the greater depths the mean current is about 85 per cent. of the surface current; and, since the bulk—over 90 per cent.—of the river's volume passes in the great depths, this percentage may be applied also to the shallower parts without material deviation from the actual discharge. Accordingly 85 per cent. of the mean surface current multiplied with the sectional area will give the volume within a couple of per cent.; but, if we should not happen to know the surface currents, the data would be of little use without a correct formula. With the Uruguay, during the periodical floods, the percentage is close upon 90 of the surface current; during the September and October floods it will exceed 95; which, combined with the greatly increased surface current, swells its volume enormously in comparison to the river's small surface current at low water; with a small sectional area, and a percentage of under 60. We mention these figures only to show, how variable the percentage of the mean current may be in the same river and at the same locality, and that the percentage should never be called a " coefficient " because it is a variable quantity; which is contrary to the meaning and the sense of the term. We may at all times correctly ascertain the volume of a river by means of the current integrator; even at irregular and exceptional localities, with which no formula could possibly deal. The volume of a river should always be ascertained by the integrator; it may be accomplished with certainty and without much trouble to within half a per cent. of the actual discharge. The volume so ascertained, will be a safe basis for further argument and consideration, and for the calculation of the various quantities which alterations and modifications in the existing channel may involve. Wherever practicable, the river's volume should be observed on numerous lines across the section by the integrator, and not calculated by any formula from given quantities, such as: sectional area, mean depth, fall per mile, &c. The "fall per mile" is the delicate quantity, difficult to ascertain without current observation; slight variations of which will materially affect the result. Indeed, the operation ought to be the reverse; from the volume and the currents, the inclination ought to be calculated. It is necessary to ascertain the fall of a river's surface between two distant points as a basis upon which the gradients of its surface may be constructed; and, although they may be ascertained by calculation, it is important to have a check by direct measurement on the accuracy of the calculation and the true outline of its longitudinal section.

Within natural channels the application of formulæ, however correct they may be, will always remain limited, and most of the quantities should be ascertained by observation and not by calculation; because a formula cannot take into proper account local disturbances, arising from all kinds of irregularities which may even escape attention. It is different with artificial channels; and these are the legitimate field for formulæ and calculation. Artificial channels are invariably regular in outline, and it is within our power to limit disturbances to any extent, and the results of correct formulæ will be as reliable and safe for argument as those of direct observation. We need both, correct observations and correct formulæ; and, it would be difficult if not impracticable in many instances to devise and to design improvements without our possessing both, practice and theory.

Much attention has been paid in the preceding chapters to the subject of inclination of a river's surface, usually expressed as its " fall per mile." The question of inclination is perhaps the most delicate of any, and it is not surprising that it should be least understood; but it is lamentable that many, who ought to know better, invariably confound the inclination of a river's surface,—that most delicate differential plane,—with the inclination of a river's bottom,—that coarse and most irregular surface;—maintaining with an air of gravity, that the one may be taken for the other; or, that the one determines the other. There exists, indeed, a distant connection between the two, as much as there is one between a mountain and its shadow; to confound the one with the other is, however, equally strange and incomprehensible.

A section may be made anywhere across a river; yet, the proper selection of the locality suitable for an instructive and useful section, causes often much embarrassment. We have seen, that bends interfere most seriously with the regular formation and action of currents; that they cause a kind of revolution against established principles and laws. The currents of a river, which had to pass a succession of sharp bends, are effectually "meaned"; order and law appear subverted and confusion to reign supreme. In such localities formulæ —the expression of order and law—can never be of any use. The movements of currents are complicated enough in regular and straight channels, and we do not believe that it would be of any practical value further to complicate them with the effect of bends, although the question may be of theoretical interest. A section should be preceded by a straight reach; and if there be none, it should be placed near the locality where the radius of bend changes from one bank to the other. The choice of a suitable locality for a useful section is always one of difficulty, and before the line is finally adopted a number of trial-soundings should be made.

The centres of gravity of section and of currents determine, whether or not the section had been well chosen, and its situation may be considered favourable. Should the two centres fall in the same vertical plane, the locality is favourable for observation. Absolute coincidence should not be expected, and it could be only accidental; but the variations referred to the width of the river, should be small. Thus, for example, the difference between the two centres in the Salto section of the Uruguay amounts to $\frac{1}{153}$ of the river's width from margin to margin; which may be practically taken as a coincidence, the variation being within the margin of possible accuracy. In the Rosario section of the Paraná the difference between the two centres amounts to $\frac{1}{313}$ part of its width; which may be considered still a very good agreement and a favourable locality; in the Palmas section the difference in the position of centres is $\frac{1}{31}$ part of the river's width; and here the disturbance of currents is getting inconvenient, and the locality cannot be considered favourable; although the radius of curvature is between 1¼ and 2 miles; the bend forming an arc of about 20 degrees.

To convey an idea of the magnitude of the rivers which have been considered and analyzed in the preceding chapters, we have shown on Plate V. several of the larger known rivers, such as the Danube and the Thames of Europe; the Mississippi of North America. They are all drawn to the same scale, and their relative size may somewhat be appreciated. The Mississippi is not unlike the Uruguay in dimensions and other features; we have similarity in width, depth, currents and fall, although the North American is the larger of the two. Comparing, however, the Paraná with the Mississippi, the former might claim the latter as his excentric daughter under fourteen. The low-water dimensions measure a river's greatness, although things of different nature and character do not bear strict comparison. What we, however, understand by "Greatness" is possessed in an exceptional degree by the Paraná.

The proper application of the principles derived from the survey of the Great Rivers cannot fail to be useful in river improvements. Circumstances may modify their application; within a river, for example, the volume is a given quantity; within an estuary it is practically unlimited. By deepening the channel of a river, a variety of effects will be produced, which may be beneficial in some respects, injurious in others. It should be our aim not only to keep the balance on the right side, but to reduce injury to nil; and, patiently considered and judiciously managed, it may in most cases be attained.

The science of River Engineering had been comparatively neglected. In no other branch had graver and more expensive blunders been committed. No doubt, it was in a great measure guesswork—matter of opinion ;—reasoning upon the solid basis of observation having been at a discount. Rivers inundate vast tracts of land and continue to do so, not because they have too much water or that their depth is not great enough, but because we are too shallow to offer the proper remedy, which in nine cases out of ten it would be in our power to give and under our control to accomplish. Channels of estuaries and harbours silt up, and their navigation becomes more and more difficult, not because the action of the sea is different to-day from what it was a century ago, but because we are, in the first place, more alive to alterations; and, in the next, because in many instances we had already aggravated the situation by desultory attempts at improvement. There is hardly a more intricate problem than the consideration of currents and their ultimate effect on the channels of an estuary ; yet, with but few exceptions, "improvements" are made without an attempt to trace the history of the estuary, much less the causes at operation ; and, we generously leave it to the next generation to reflect over and learn by our present mistakes. Solid improvement is of slow growth ; and we should proceed with patience and perseverance. The field of engineering is wide, unlimited—and there is room for all who join with a determination to labour with devotion—faithfully to uphold the truth of our science. The secret of genuine success consists in the discovery and the application of its truths. Those, endowed with genius, unconsciously perceive it ;—others, will discover it by degrees and may practise it with equal success. All may acquire the principles of the science, and by their conscientious application confer untold benefit on society of which they form part themselves. The truths, which the survey of the Great Rivers disclosed and brought to light, will, we doubt not, do some good and advance the science of engineering ; and the sooner they are perceived, understood, and applied, the prompter will be the advantages which their practice will ensure to engineers of our generation.

APPENDIX.

# APPENDIX.

# APPENDIX.

## THE IMPROVED CURRENT METER, AND ITS APPLICATION.

A GOOD and reliable instrument for the measurement of the velocity of currents of rivers or of estuaries, is of primary importance to ensure valuable observations. The records of such instruments should be the basis for calculation and argument. To avoid repetition we refer to pp. 14 and 15 of the second chapter descriptive of the instrument or " Current Meter " which had been used in the survey of the Great Rivers; and to pp. 17, 18, 19, 20, 21, explaining the use of the meter, to integrate currents, or to observe velocities at fixed levels, giving the mode of proceeding and pointing out the precautions which should be taken to ensure accurate and reliable observations. We recommend the careful perusal of the whole subject under "Currents of River" of Chapter II., and, with these premises, we proceed to give a description of the improved meter as shown on Plate VIII.

Fig. 1 shows the meter in elevation and partially in section, drawn half full size. The axis $a$ of the screw is attached by means of the boss $b$ to the spherical boss $c$ of the screw, which holds the blades $d, d$. The screw of the meter is not unlike a miniature "Griffith," though the object to be gained is very different. The diameter of the spherical boss $c$ is so determined that it will displace just as much water as to weight, as will balance the weight of the spindle $a$, the bosses $b$ and $c$ and the blades $d$; in short, the weight of all the parts which are fixed on the spindle $a$. The boss $c$ is water-tight and, when the apparatus is submerged, the screw revolves on its spindle without any pressure on the bearing $e$, inasmuch as the sphere $c$ floats all the parts, which under water weigh nothing. We have, consequently, eliminated the friction of the spindle of the screw on its bearing $e$ whilst revolving submerged under water; it is one of the main sources of the slip of the screw, and its irregularity due to increase or decrease of that friction, arising from more or less turbid or gritty water.

On the axis $a$ we have the short double thread $f$ which works in two worm-wheels $g$ and $h$. The worm-wheels revolve on a common axis $i$, which is also the centre, and forms part, of a disk $k$; which may be a little raised or lowered by the pin $l$, so as to throw the worm-wheels into and out of gear. In the drawing they are shown in gear, and as the axis $a$ of the screw revolves, so will both wheels turn round their common centre $i$. A coil-spring, round the pin $l$, always presses the disk $k$ down, tending to throw the wheels $g$ and $h$ out of gear. By the aid of a nut $m$ the pin may be screwed up, and the wheels thrown and held permanently in gear; in opposition to the coil-spring. When the nut $m$ releases the pin $l$, and allows the coil-spring to press the disk $k$ and with it the wheels away from the thread $f$, throwing the latter out of gear,—the worm-wheels immediately come in contact with another flat spring $n$, which checks their accidental movement in any direction. All

x 2

156                                         APPENDIX.

the parts previously mentioned are within the frame $o$ of the apparatus, and covered by a
stout glass plate $p$ held in its place by the ring $r$. The glass plate does not quite touch
the worm-wheels to within about $\frac{1}{50}$th inch, the small spring $s$ gently holding the wheels
towards their disk $k$. The frame $o$ of the apparatus has a projecting arm $t$ with a long
cylindrical bearing $u$, which freely revolves on a hollow spindle $v$, and the whole apparatus
may accordingly revolve on the spindle $v$. A directing vane is attached to the frame of
the apparatus at $w$, in the prolongation of the axis $a$ of the meter's screw. Another vane
is attached at $x$, at right angles with the directing vane ; the object of the latter being to
allow the whole apparatus to touch the bottom, without interfering with the free and easy
revolution of the meter's screw ; the latter vane, projecting under the blades $d$, will prevent
their contact with the ground. The whole apparatus in position for observations at fixed
levels or for current integration is shown at A, Fig. 2, for depths from 10 to 200 feet or
more ; for depths under 10 feet the apparatus as shown at A may be attached to a rod,
see Fig. 3.

The general arrangement of the meter as shown, Fig. 1, will explain itself from the
drawing. It has already been remarked, that the working parts of the meter are confined
within its frame $o$ and covered by a stout glass plate, enclosing its parts almost air-tight
under cover. Through the glass plate the divisions on the worm-wheels $h$ and $g$ may be
seen and read off by the indexes, one of which is a fine line on the glass plate to read the
divisions of the wheel $g$ ; the other index is attached to the latter wheel, under figure "100,"
and by which the relative movement of the other marked $h$ may be readily determined.
There are 100 divisions on wheel $g$ ; for each revolution of the meter's screw it will move
one division in reference to the line on the glass plate ; and, for 100 revolutions of the
screw it will make one complete turn ; the same division line of wheel $g$ returning to
the line on the glass plate. On the second wheel $h$ there are 49 divisions, the index on
wheel $g$ under the figure "100" shifting the distance of one for a complete revolution of $g$.
The value of one division of $h$ therefore corresponds to 100 revolutions of the screw ;—
the number of the hundreds is accordingly read off from wheel $h$, and the units up to a
hundred from wheel $g$.

Since, however, large divisions would require inconveniently large teeth in the worm-
wheels, if their numbers were similar to the divisions, there are on wheel $g$ twice as many
teeth as the number of divisions, viz. 200 ; and on wheel $h$ there are four times as many,
viz. 196. To make, however, the thread on the axis $a$ cause wheel $g$ to move one division
for one revolution of the meter's screw, it must be a double thread, viz. of a pitch twice
the distance from tooth to tooth.

It had also been found, that the wire used for throwing the meter in and out of gear,
should pass through the centre of the axis, round which the meter turns in obedience to the
directing vane in its rear, which always keeps the axis of the screw in the line of current ;
otherwise in gentle currents, by pulling the wire hard, the axis may be turned obliquely to
the current by many degrees, impairing the accuracy of the result.

The diameter of the screw across the blades is about 6 inches ; there are two blades of
German silver with a pitch of about 9 inches. For very strong currents, which might

injure the comparatively long and thin blades, we use short ones of steel, and these may be very thin and plane; since the variation of pitch in the short distance would not be material. Their value per revolution is different, and it should be ascertained by direct experiment.

To prevent turbid and gritty water getting to the mechanism under the glass plate, thereby increasing friction and introducing possible irregularity of movement, the whole space under the glass plate should be filled with clear and pure well-water by holes at $x$, afterwards closed with their screws; and it should be done immediately preceding an intended observation. To prevent pressure forcing in the glass plate, by rapid lowering into great depths,—in case air-bubbles had been left behind not completely filling all cavities under the cover,—there is a fine channel $y$ for communication between the enclosed chamber and the outside, which will always establish similarity of pressure in and outside the apparatus. After completion of the observations, the water may be let out at $x$, and the spindle $a$ may be taken out, wiped and oiled, and again replaced.

There are two modes for observing currents; either, first, by putting the worm-wheels into gear with the meter's screw, by tightening the nut $m$, and then observing the number of revolutions for the time the meter was immersed and exposed to the current; or, second, by first immersing the meter, and then by pulling and letting go a wire to throw the meter into and out of gear at given times, whilst it is submerged. The first mode is not practicable at fixed levels; the second is practicable for all except for current integration, which should be done by the first mode. To work the instrument for fixed levels,—be it at the surface or any distance below it,—the nut $m$ should be first loosened; the spiral spring will then throw the worm-wheels out of gear, and the meter's screw may revolve whilst the apparatus is lowered to the desired level without registering revolutions. A wire having been attached to the eye of the pin $l$, and passed through the hollow spindle $v$ out at the aperture $x$, reaching to the observer's hand, it is drawn tight the moment the observation is to begin; the instant it is to end, the wire is let loose, throwing accordingly the wheels in and out of gear. The position of the indexes should be read and booked before and after each observation; the difference of their position will correspond to the time of the trial.

For surface current observations the complete apparatus may be screwed to a wooden staff about 3 inches wide and ½ inch thick, as shown in Fig. 3. The observation may be made either with or without the wire, whichever may be more convenient; in the first mode the meter is immersed at a given signal: usually the centre of screw is placed 12 inches below the surface, and having exposed it to the action of the current for the time intended, say one, five, or more minutes, it is at another signal raised above the surface; the moment the blades appear above water, their revolution should be checked by an assistant. Signals are not necessary; for, the observer may himself look at the watch, holding the staff of the meter in hand, and immerse or raise it at the time intended for observation. Should the trial be made by wire, a sailor may hold the staff and meter in position, and the observer may draw tight or let go the wire at his discretion, noting the exact time of commencement and termination of trial to the fraction of a second. On every observation there should be at least one check trial, which should agree within the margin of accuracy, otherwise the observation is incomplete or doubtful.

For observations at fixed levels, say 10, 20, 30, 40, &c., feet below surface, the meter arranged in the manner shown on Fig. 2 may be lowered into position as shown, Fig. 4. A boat *A* having, according to depth, been moored 50 or 100 yards ahead of the section by double moorings to prevent oscillations, another boat *O* is attached to the former by a line *l*. A third boat *B* may be moored a similar distance from *O* by double moorings; all the boats to be in the line of current, and the boat *B* should be connected with *O* by another rope *m*; both lines *l* and *m* having been hauled tight, observations at fixed levels may be resumed at *O*. The apparatus as shown, Fig. 2, should then be lowered from the observatory *O* by two light chains attached to the straps at *r* and *s*, the chains being marked every 5 feet; and a line *p* having been attached to the boat *A* and to the eye *x* of the staff, and another line *q* from the eye *y* to the boat *B*. The lines *p*, *q*, may be hauled tight, and the former should be fastened permanently to the boat *A*, whilst the latter may be paid out as the whole apparatus is being lowered from *O* to the desired level. The operation requires at least four sailors. The apparatus having been lowered to the desired level, as ascertained by the marks on the chains, and holding the staffs *t, t*, horizontally, the wire of the meter may be drawn tight for the commencement, and let go for the termination, of the observation, noting the exact time; and by raising the apparatus to the surface, the indexes may be read and the number of revolutions of the meter's screw ascertained by comparison with their former position. The apparatus may then again be lowered to the same level for the check trial. Observations at fixed levels require the greatest care, and are more difficult than others, and the crew should be trained. Should the currents be changeable, like those of estuaries, the check observation should be made at the surface, including the trial at the low level between two or more at the surface. For observation of currents at the bottom of a river or an estuary, the apparatus may be lowered until the disks *u* touch the bottom, which will bring the blades of the meter's screw within an inch of it; should it be desirable to ascertain the current 6, 9, or 12 inches above the bottom, the staff *t, t*, of the apparatus may be pinned to any of the holes in the straps *r, s*, and so ensure the distance while the disks *u* are resting on the bottom.

To integrate currents from surface to bottom—the most important of all observations—the third boat *B* is not wanted; the apparatus may be lowered to the bottom and raised to the surface without the line *q*; but the boat *O* should be moored by two lines to prevent oscillations. For these trials no wire is wanted, and the worm-wheels are thrown into gear before the commencement of trial, the meter commencing to register the moment it is immersed in water. The apparatus should never be held stationary at any level, but from its immersion constantly be on the move, and it should be lowered and raised at a uniform rate. To ensure that the staff *t, t*, is held horizontally during the integration of currents, red and white marks should be attached alternately on each chain at equal distances, one sailor calling out the colour of the mark reached, with which the other sailor should keep pace, so that red and white colours are immersed nearly simultaneously; otherwise the staff *t* would soon be at 45 or 90 degrees with the horizon, entirely vitiating the trial; there being no time to read the figures on the chain. It is immaterial how fast or how slowly the apparatus may be raised or lowered, as long as the rate remains the same during the trial. Nor does it matter how often the distance from surface to bottom had been passed over. For shallow depths under 20 feet it should be done several times in succession before the meter is finally raised out of the water; for the main trial three to six times, for the check observation once or twice up and down. For greater depths, from 20 to 100 feet, the main trial may be twice up and down; the check observation once over the distance from surface to bottom and back to surface, which may be considered as one run. It is difficult to make many runs in great depths without special arrangements to lower and raise the whole apparatus; it exhausts the raising power of the sailors, which may cause irregularity of movement. Surface current trials should immediately precede and succeed an observation for the integration of currents. The bottom current may also be observed from $O$ without the third boat $B$, but currents at intermediate levels are not safely observed without the third boat and the line $q$, because, especially in weak currents, the whole apparatus may oscillate to and fro on the line of current like a pendulum. It is of little consequence with observations extending over five or ten minutes, because they eliminate one another; but with minute observations it may seriously affect the result.

For greater depths than 100 feet the spiral spring of the meter must be of special strength, because it takes a pull of many pounds before even a thin wire will nearly come into line; the current bending it considerably, and the additional pull should at a given moment throw the meter into gear. Moreover, the weight of the disks $u$ should be greater, otherwise the chains are much bent by the current, and the vertical distances become erroneous; all this, however, affects only current observations at fixed levels.

The position of the indexes of the meter should always be noted in the Survey Book, as corresponding to the commencement and the termination of a trial, and they should never be copied from one observation to the other. Since the indexes are not changed by hand, the reading at the termination of a trial is usually the same as that for the commencement of the next; sometimes there are accidental alterations; the rule should be, always to read anew, and to book the position of indexes as a check on former readings. An error in the readings causes much confusion, and the loss of two trials if they be copied, and the termination of the one be taken as the commencement of the next.

It will be seen, that the improvements in the construction of the meter as above described, reduce the friction of its mechanism to a minimum, next to nothing; the resistance to the movement of the spindle of the screw is only due to that between a liquid and a solid at low velocity; and the movement of the mechanism under the glass cover is slight in comparison to that of the screw; and, with good workmanship, the friction is hardly perceptible. The great advantage of the arrangement, however, consists not so much in the extreme ease with which all parts of the meter may be set in motion, accurately recording the gentlest currents, but in the circumstance that the resistance of the mechanism remains always the same, no matter whether the meter be tried in clear, turbid, or in gritty water; because its mechanism is always immersed in the same medium, with which the space under the glass cover had been filled previous to trial; it may be pure water; oil causes verdigris; we recommend pure water, because it may always readily be obtained in the small quantity required by filtering it through blotting-paper.

The improved current meter holds in reference to the ordinary meter, of similar construction in many respects, the same position as a chronometer would to a common Geneva watch; both may do good service according to the object and the requirements, and we ought to be familiar with a common watch before we try to handle a chronometer. Without some practice neither can be properly managed as a matter of business, and we ought to make first a number of trials, to obtain practice and to get acquainted with the instrument, before we engage seriously upon observations. A pond or a canal will offer many advantages to obtain practice, by moving the meter at certain velocities through still water, and it will offer opportunities to check the accuracy of the observations through the distances traversed by the meter; and, having obtained some practice as shown by the coincidence of results, we may also determine the value of one revolution of the meter's screw at different velocities, which should never be neglected; it is as important as the "rate" of a chronometer; although the makers may give it, it may change, and must be checked and ascertained from time to time; a slight bend of the screw blades will alter the value a little, which should be accurately known by periodical trials and examination. For very important observations there ought to be two meters; all the trials should be made by one. The second meter should be only used occasionally, to check the value of the first. With such current meters we may determine with as much certainty the movement of water, as a captain by the aid of his chronometer may determine the position of his vessel anywhere on the globe. A hydraulic engineer is as much in need of a couple of good current meters, as a captain navigating the wide ocean is in need of a couple of good chronometers. The improved current meter may be useful to hydraulic engineers; and, to facilitate its general adoption, we have, on consideration, determined that its manufacture should be free; the engravings show all the details for its successful construction. We commend the splendid engravings of the Plates of this work by Mr. Thomas Kell; they are specimens of the art of lithography; and, although we consider their execution above praise, we cannot omit to acknowledge our satisfaction. We observe, that the eminent mathematical instrument makers, Messrs. Elliott Bros., of London, made our meters, and are acquainted with all requirements: they charge about ten guineas for the complete instrument.

## EQUATION OF METER.

By "equation of meter" an algebraic expression is understood, which will, by calculation, give the velocity of current in feet from the number of revolutions of the screw per minute. If no friction whatever existed in the mechanism of a meter, the number of revolutions observed would be proportional to the number of feet; if, for example, the pitch of the screw were such, that a current of 1 foot velocity per minute or per second, impinging on its blades, would produce one revolution, then a current of 10 feet velocity would in the same space of time produce ten revolutions; of 100 feet, 100 revolutions, and so on. In that case we should have only to ascertain the value of one revolution of our instrument, expressed in feet, and whatever it may be, if multiplied by the number of the observed revolutions, we should obtain the corresponding velocity in feet. The equation of the meter would then be that of a straight line passing through the origin of the co-ordinates, viz. $y = a x$, in which $y$ may represent the number of revolutions of the screw, and $x$ the corresponding velocity in feet, and $a$ the tangent of inclination, or the coefficient of the particular meter. There is, however, always some friction, which to overcome there must be a difference of velocity between the current and the blades of the screw; and this difference, nearly a constant quantity, bears a variable proportion to the actual current; thus, for example, if the necessary velocity of current to overcome that friction were 10 feet per minute with a particular meter, then a current which would produce 100 revolutions with an instrument without friction, could only produce 90 revolutions with the particular meter; a current of 50 revolutions could only register 40, and one of 10 revolutions would register *nil*; the difference of 10 revolutions being consumed by friction, representing the slip of the screw. Although the " difference " necessary to overcome the friction is sensibly a constant quantity for all meters, it **bears a** very different percentage if referred to various currents; in the example above quoted it would have been 10 per cent. with first; 20 with the second, and 100 with the third velocity of current. As long as the " difference " will remain sensibly a constant quantity, a straight line would nevertheless represent the dependence between revolutions of the screw and the number of feet of the current, but it would intersect **the** abscissæ a certain distance from the origin of the co-ordinates. To **determine the** position of the straight line in reference to the co-ordinates, two observations **are necessary under** different velocities, both of which the meter **must register with accuracy. A number of** trials should be made **at a low velocity**; another number of observations **at a high velocity ;** the **result** of the two sets of trials will determine two points, which will **fix the position of** the straight **line.**

An example will illustrate the proceedings. We will take current meter No. 1 with which all observations on the Great Rivers had been made. Its value had been determined on reservoirs. The distance through which the meter passed while immersed was 189 feet 9 inches, at a velocity of about 76 feet per minute (the base being passed over in 2m. 30s.), the average of three trials gave for the base 172 revolutions, or for a distance of 75·90 feet passed over in one minute 68·80 revolutions; in other words : for a current of 75·90 feet per minute the meter registered 68·80 revolutions. In a similar manner, the base being passed over on the average 1m. 2s., the meter registered 182 revolutions for the distance, or for a velocity of 183·64 feet per minute the number of revolutions had been 176·13 ; a current of 183·64 feet per minute made **the** meter revolve 176·13 times **per** minute. **We** have, therefore, for two different velocities their corresponding **revolutions. We** will **give the** trials themselves **to** illustrate **the proceedings.**

TRIAL OF CURRENT METER No. 1. 19TH NOVEMBER, 1870. WATERWORKS RESERVOIR, BUENOS AYRES.

| No. of Trial. | Distance of Run in Feet and Inches. | | Time. | | No. of Revolutions of Meter for Distance. | Remarks. |
|---|---|---|---|---|---|---|
| | ft. | in. | min. | sec. | | |
| | 1st Series. | | Low Velocity. | | | |
| No. 1 | 189 | 9 | 2 | 36 | 170 | Calm weather. |
| „ 2 | 189 | 9 | 2 | 33 | 172 | |
| „ 3 | 189 | 9 | 2 | 21 | 174 | |
| | 2nd Series. | | High Velocity. | | | |
| „ 4 | 189 | 9 | 1 | 0 | 182 | |
| „ 5 | 189 | 9 | 1 | 2 | 182 | |
| „ 6 | 189 | 9 | 1 | 4 | 181 | |

Many more trials had been made, of these three had been taken in which the time of passage over the distance, marked off on shore by poles, had been nearly alike, and these were grouped together. It is more difficult to ensure uniformity of movement at low than at high velocity ; and trials in which the movement had been irregular should be rejected.

The above trials determine the ordinate and the abscissæ of a point of the line at low

Y

velocity, and of another point at high velocity; and if we call the co-ordinates of low velocity $y$ and $x$, and those of the other point due to high velocity $y_1$ and $x_1$, we have the following two general equations of the straight line passing through the two points, viz. :—

$$\left. \begin{array}{l} y = ax + b \\ y_1 = ax_1 + b \end{array} \right\}$$

from which $a$ the tangent of inclination of the line, and $b$ the constant quantity, may easily be determined, viz. :—

$$\frac{y_1 - y}{x_1 - x} = a, \text{ and } \frac{(y_1 + y) - a(x_1 + x)}{2} = b$$

and substituting in the above equations the value of $y$ $y_1$, $x$ $x_1$, determined by the experiment, viz. :—

$$y = 68 \cdot 80, \quad y_1 = 176 \cdot 13, \quad x = 75 \cdot 90, \quad x_1 = 183 \cdot 64,$$

we obtain $a = 0 \cdot 9962$, $b = -6 \cdot 811$, or the equation of the line will be :—

$$y = 0 \cdot 9962 x - 6 \cdot 811.$$

Since, however, an observation always gives the number of revolutions per minute ($y$), and we want to find the corresponding velocity ($x$) in feet per minute, we may resolve the equation for velocity, viz. :—

$$x = 1 \cdot 00381 y + 6 \cdot 837 \quad \ldots \quad (1)$$

Equation No. (1) had been used for all observations on the Great Rivers, to convert the revolutions of the screw per minute into the equivalent velocity of current in feet per minute.

Example 1st.

Revolutions of meter per minute, observed = 100.

$$\text{Current} = \left\{ \begin{array}{l} 100 \cdot 381 \\ + 6 \cdot 837 \end{array} \right\} = 107 \cdot 218 \text{ feet per minute.}$$

Example 2nd.

Revolutions of meter per minute, observed = 1.

$$\text{Current} = \left\{ \begin{array}{l} 1 \cdot 0038 \\ 6 \cdot 8370 \end{array} \right\} = 7 \cdot 841 \text{ feet per minute.}$$

Example 3rd.

Revolutions of meter per minute, observed = 0.

$$\text{Current} = \left\{ \begin{array}{l} 0 \cdot 000 \\ 6 \cdot 837 \end{array} \right\} = 6 \cdot 837 \text{ feet per minute.}$$

It appears from the third example that current meter No. 1 ceases to register or to indicate all currents under about 7 feet per minute; the slight pressure of so small a current would not be sufficient to overcome the friction of its mechanism. As a matter of fact, the instrument does not register currents under 9 to 10 feet per minute, or about 2 inches per second velocity; from which point it begins to operate, although it should never be used so near its margin of possible operation. Strictly considered, the line representing the dependence between revolutions and currents would not be straight, but assume a parabolic form, especially near the abscissæ; it is, however, not desirable to rely on observations which come near the margin of the instrument's power; and accordingly, it is of no practical value to complicate the equation with such considerations. It should, however, be understood that the accuracy of the meter is by no means limited to 10 feet per minute or about

2 inches per second, but that the apparatus commences to register from that small velocity ; its records, however, for ordinary velocities, are correct within 1 inch per minute, a small fraction of an inch per second,—so small, that we know of no other means to trace it.

For ordinary purposes, the revolutions of the meter may also be converted into velocity of current by geometrical construction, without using the equation of the meter in its algebraic form. Suppose O X and O Y, Fig. 5, Plate VIII., to represent the line of abscissæ and of ordinates, we may by a convenient but arbitrary scale plot the result of the experiments and find the points by construction; one for low, and the other for high velocity. The diagram, Fig. 5, is shown on a small scale to save space; 1 inch is assumed to represent on O Y 100 revolutions, on O X 100 feet per minute. In practice the scale ought to be about ten times as large, or 1 inch to represent ten revolutions or 10 feet velocity, in order to be able to read the distances with accuracy by the scales. If, by the scale of the diagram we make the distance $O q = 75 \cdot 90$ feet $= x$, and $n q = 68 \cdot 80$ revolutions $= y$, we obtain a point $n$ of the line; if we further make $O p = 183 \cdot 64 = x$, and $p n = 176 \cdot 13 = y_1$, we obtain another point $m$; and by connecting $m$ with $n$, and prolonging the line in both directions, we obtain the line $l\,l$, which represents the dependence between revolutions of the meter's screw and the velocity of current in feet per minute. If, for example, we make $O r = 200$ revolutions by the scale, and from the point $r$ draw a line parallel to O X until it intersects the line $l\,l$ at $s$, the distance $r\,s$ will represent the velocity of the corresponding current, which by the scale of the diagram would be 207 feet 7 inches; and so on for any number of revolutions the corresponding velocity may be obtained by construction alone. We know from the equation of the meter that for $y = o$, $x$ would be $6 \cdot 837$ feet; that is to say, the line $l\,l$ intersects the abscissæ at a distance of 6 feet 10 inches from O, which will be a check on the accuracy of the construction. These examples refer to meter No. 1 of the Great River Surveys; every meter has its own equation, to be determined by experiment.

LONDON : PRINTED BY WILLIAM CLOWES AND SONS, STAMFORD STREET AND CHARING CROSS.

Nº 1.

CHART

OF THE

LA PLATA

AND THE

DELTA OF THE PARANÁ.

Soundings in Fathoms.

1873.

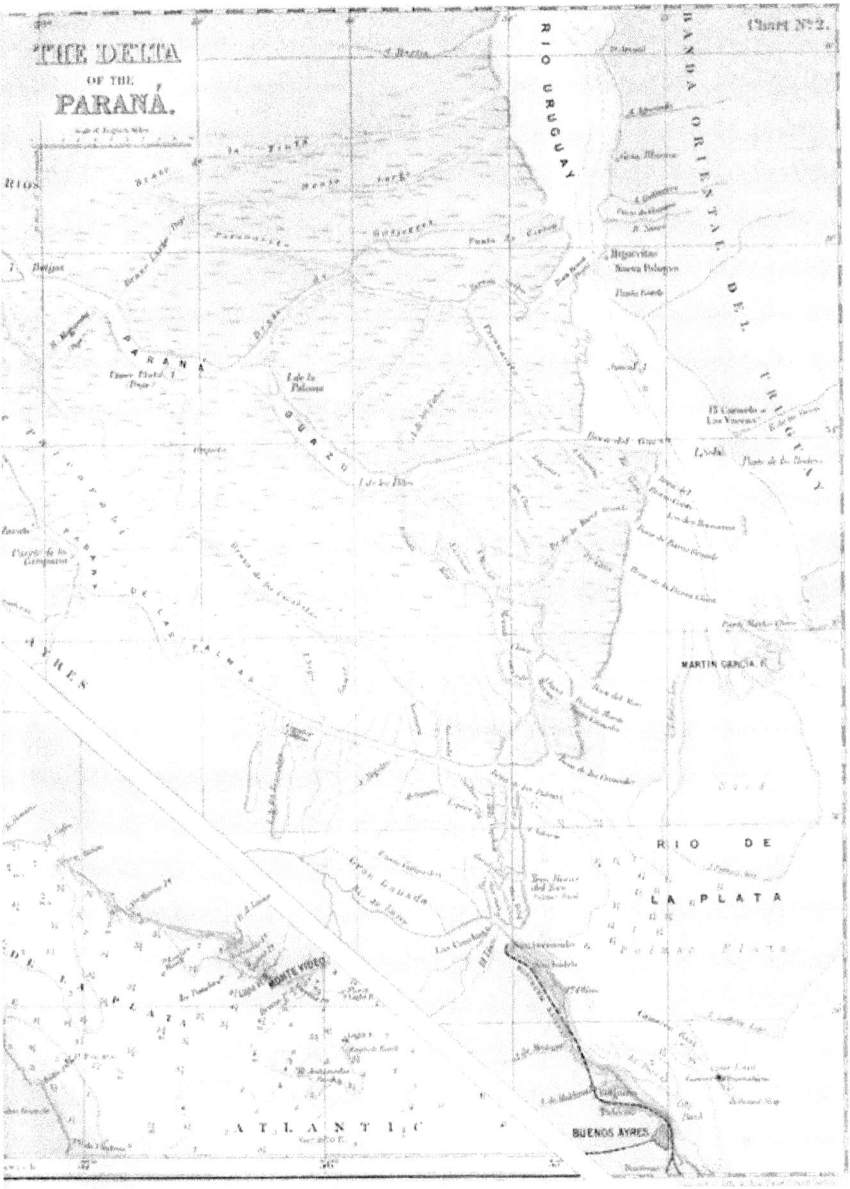

Chart No. 2.
THE DELTA OF THE PARANÁ.
RIO URUGUAY
BANDA ORIENTAL DEL URUGUAY
PARANA
GUAZU
PARANA DE LAS PALMAS
MARTIN GARCIA I.
RIO DE LA PLATA
MONTE VIDEO
ATLANTIC
BUENOS AYRES

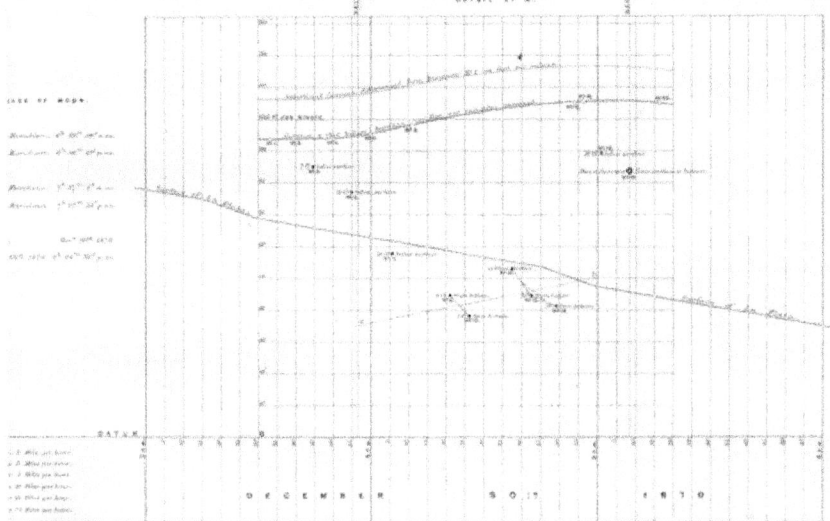

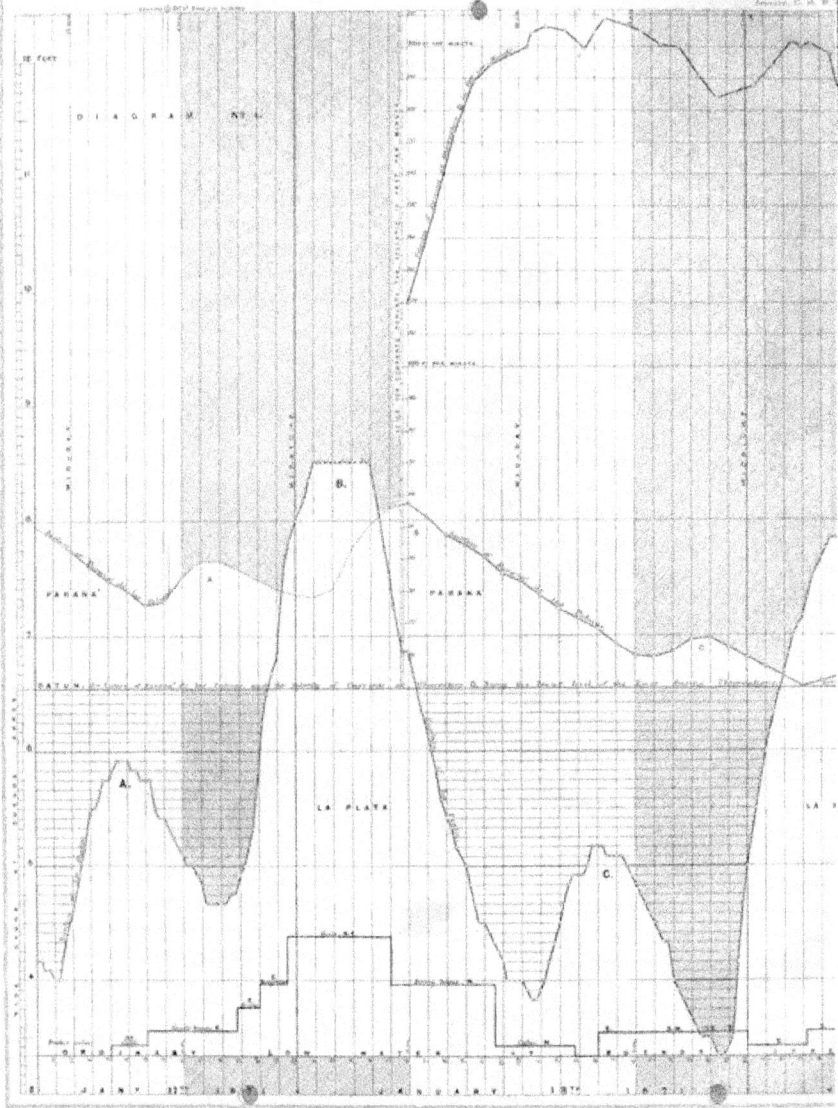

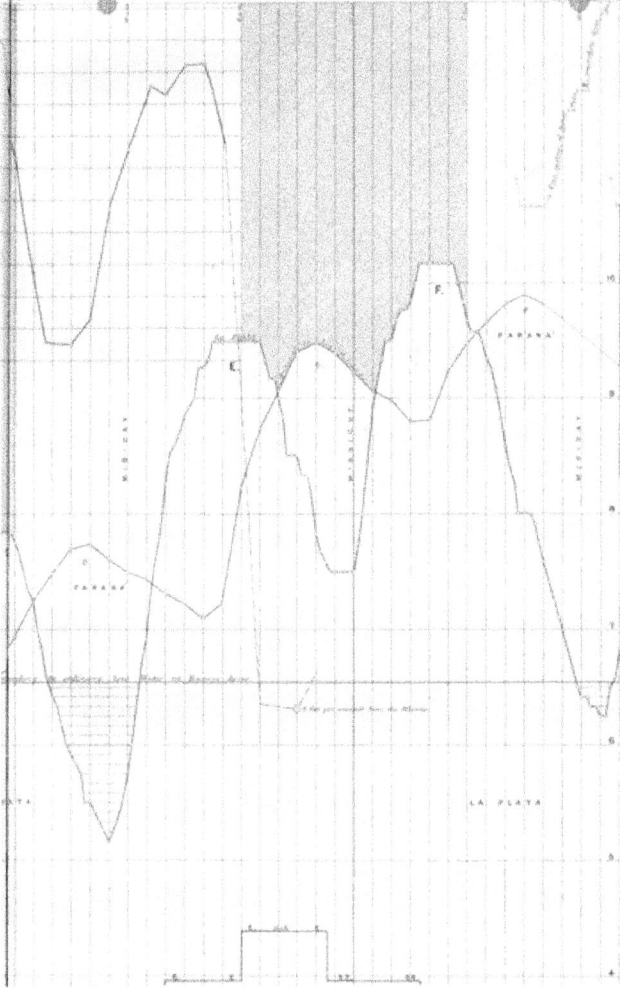

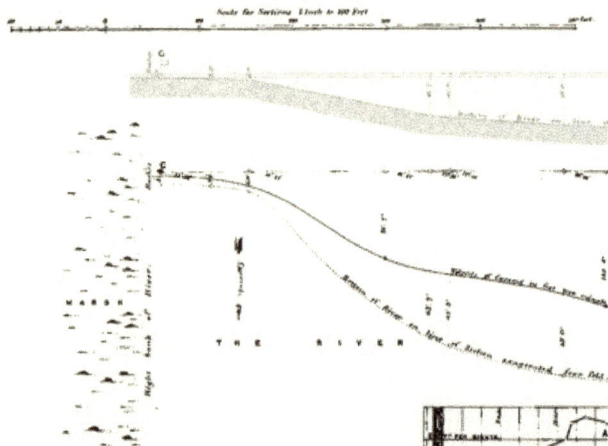

Scale for Sections 1 inch to 100 Feet

MARSH

THE RIVER

MEMORANDA.

DETAIL AT A.

Diagram shewing Superficial Velocity and Velocity 3 Feet below Surface at 4.15 p.m. January 18th, 1851.

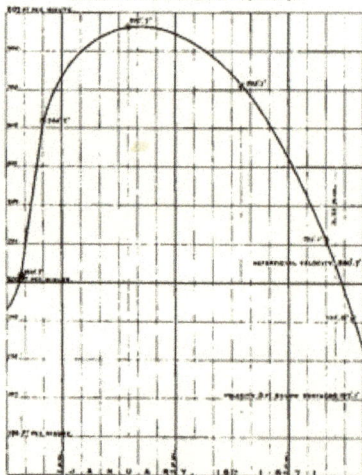

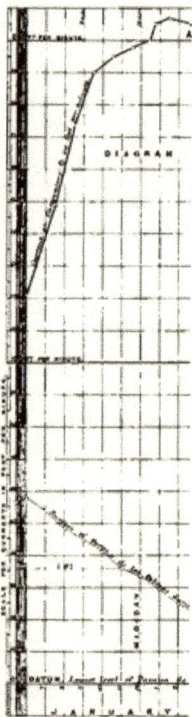

DIAGRAM

JANUARY

JANUARY

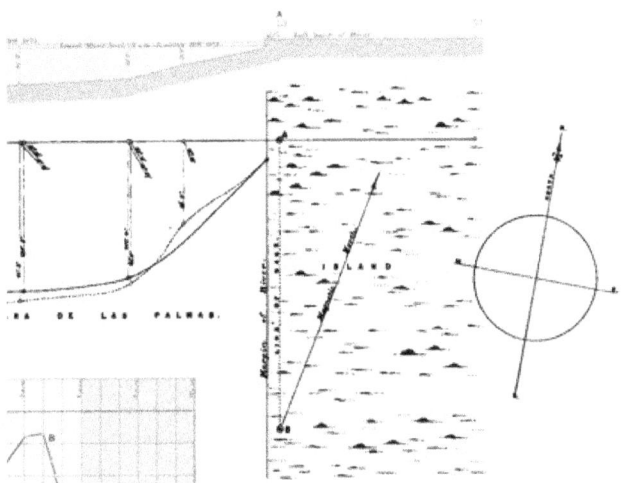

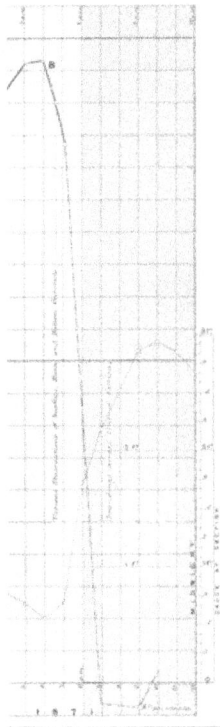

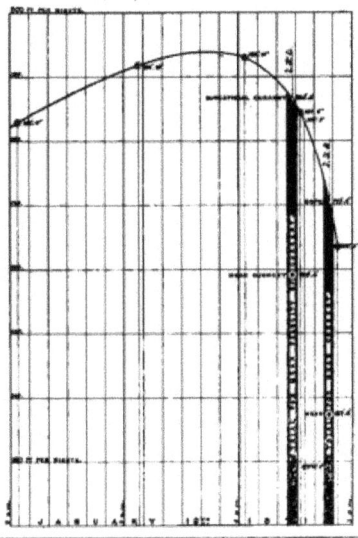

RA DE LAS PALMAS.

DETAIL AT B.

Diagram showing Mean Currents at 2.30 & ... p.m. and at 9.30 & ... p.m.
with corresponding Surface Velocities. January, 18th, 1871.

J A N U A R Y    1 8 7 1

Thomas Kell, Lith. 40 King Street, Covent Garden.

# GREAT

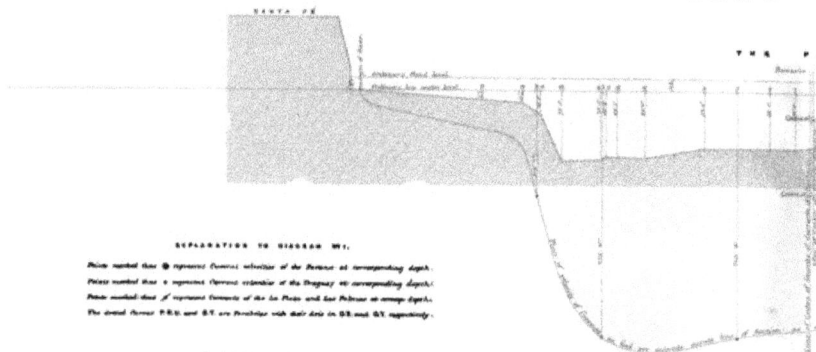

## EXPLANATION TO DIAGRAM Nº 1.

Points marked thus ⊕ represent Central velocities of the Parâna at corresponding depth.

Points marked thus ○ represent Current velocities of the Uruguay at corresponding depth.

Points marked thus ∫ represent Currents of the La Plata and Los Palmas at average depth.

The dotted curves P.R.S. and S.T. are Parabolas with their Axis in D.S. and G.T. respectively.

## DIAGRAM Nº 1.

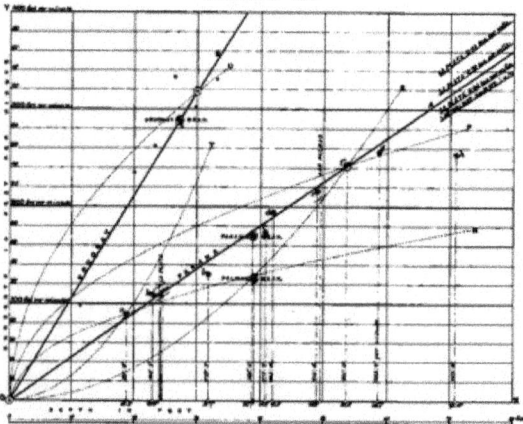

DEPTH IN FEET

## DIAGRAM Nº 2.

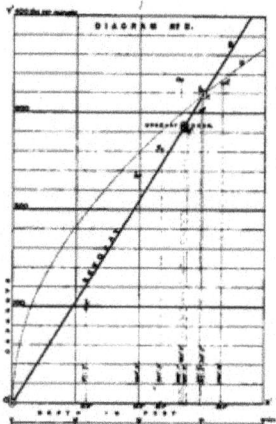

DEPTH IN FEET

## THE MISSISSIPPI AT VICKSBURG.

### Section Nº 15.

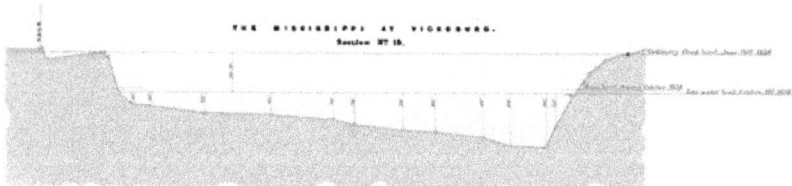

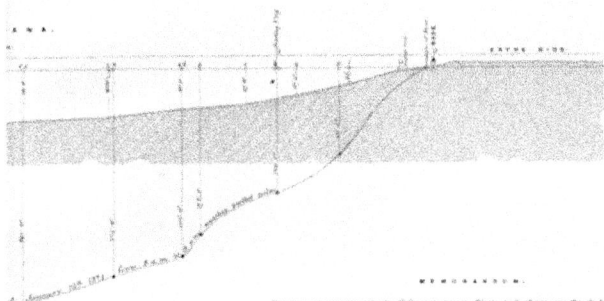

THE HIPP.

## MEMORANDUM.

THE URUGUAY.

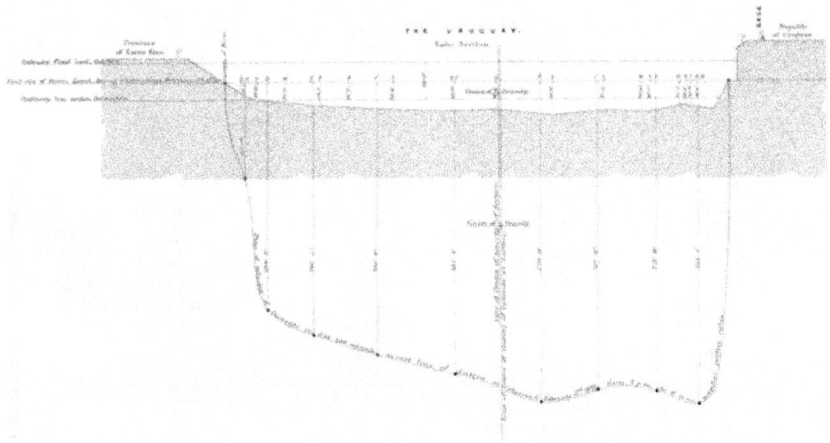

Republic
of Uruguay

THE THAMES

THE DANUBE AT VIENNA.

## EXPLANATION.

The Seawave Section had been drawn in three divisions to keep the size of the Chart within moderate limits. The three divisions united at W. W. and C. whole Section across the Passage. The figures above low water line denote the distance of the Point from the line of Base at A. and the figures below water line denote the depth in feet and inches at low water, or they denote the current in feet and inches per minute. Point the range is 1247 C. line from A. with 88 ft. line depth of water, and Point b. is 3482 C. line from A. with a current of 923 ft. 30 ins. per min.

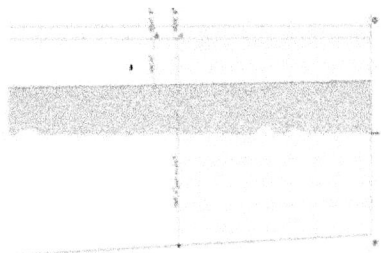

**EXPLANATION.**

The Entire Section had been drawn in two divisions to keep the size of the Sheet within convenient limits. The two divisions united at B.B. and B.B. form the whole Section across the Progaony. The figures above "Level-during Observations" denote the distances of the Point from the line of Dam at A. and the figures below that level denote the depth in feet and inches at the time of Observations, or they denote the Current in feet and inches per minute. Thus, for example, Point A. is 4085 ft from A. with 20 ft 4 ins depth of Water, and Point B. is 7869 ft ins. from A. with a Current of 201 ft 4 ins per minute.

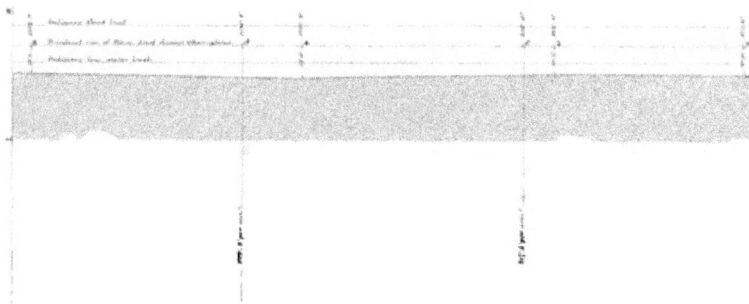

# R I V E R S .

U R U G A Y .

SECTION.

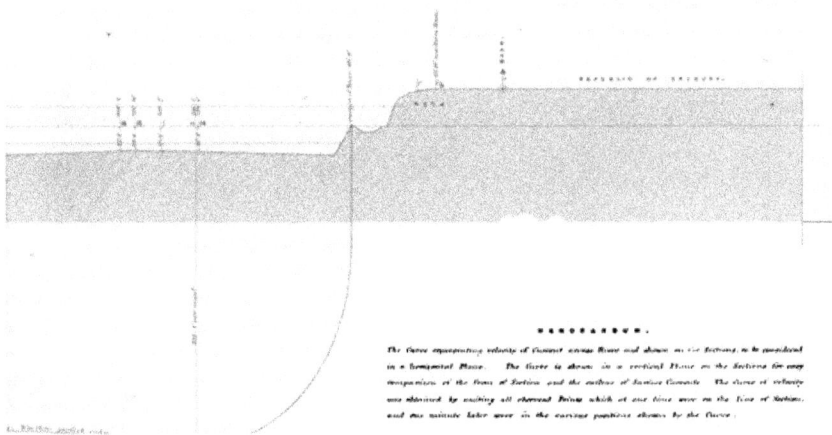

MEMORANDUM.

The Curve representing velocity of Current across River and shown on the Sections, is to be considered in a horizontal Plane. The Curve is shown in a vertical Plane on the Sections for easy comparison of the lines of Section and the outline of Surface Currents. The Curve of velocity was obtained by noting all observed Points which at one time were on the line of Section, and one minute later were in the various positions shown by the Curve.

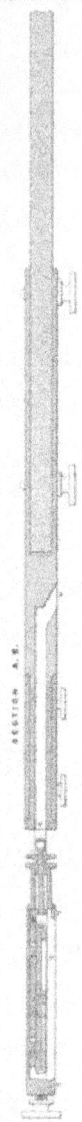

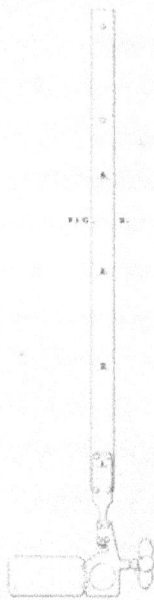

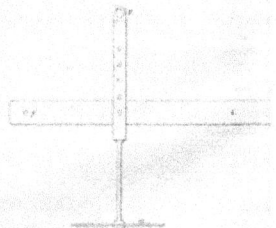

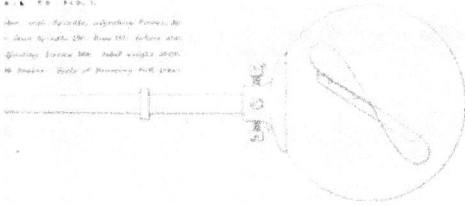

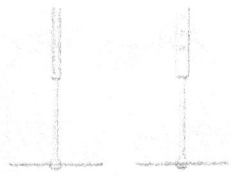

www.ingramcontent.com/pod-product-compliance
Lightning Source LLC
Chambersburg PA
CBHW021706210326
41599CB00013B/1539